W9-CEG-323

JOURNEY TO THE ANTS

JOURNEY TO THE ANTS

A Story of Scientific Exploration

Bert Hölldobler

and

Edward O. Wilson

The Belknap Press of
Harvard University Press
Cambridge, Massachusetts
London, England
1994

Printed in the United States of America

SECOND PRINTING, 1994

Designed by Marianne Perlak

This book is printed on acid-free paper, and its binding materials have been chosen for strength and durability.

Library of Congress Cataloging-in-Publication Data
Hölldobler, Bert, 1936–
Journey to the ants: a story of scientific exploration / Bert Hölldobler and Edward O. Wilson.
p. cm.
Includes index.
ISBN 0-674-48525-4
1. Ants. 2. Insect societies. 3. Hölldobler, Bert, 1936– . 4. Wilson, Edward Osborne, 1929– . 5. Ants—Research. I. Wilson, Edward Osborne, 1929– . II. Title.
QL568.F7H575 1994
595.79'6—dc20

94-13386
CIP

For
Friederike Hölldobler
and
Renee Wilson

Contents

Preface

The Ants, the monograph we published in 1990, met with critical success and surprisingly wide public attention. But it is a technical book, aimed primarily at other biologists, as well as an encyclopedia and vade mecum of myrmecology, the scientific study of ants. And because exhaustive coverage is its primary aim, it is outsized, containing 732 pages of tables, figures, and double-columned text, measuring 26 by 31 centimeters in hard cover, and weighing 3.4 kilograms. *The Ants,* in short, is not a book one casually purchases and reads cover to cover. Nor does it try to convey in any direct manner the adventure of research on these astonishing insects.

Journey to the Ants condenses the best of myrmecology to a more manageable length, with less technical language and with an admitted and unavoidable bias toward those topics and species on which we have personally worked. Where special terms must be used because of the specialized nature of the subject, we define them on the spot.

Our approach is thematic at the beginning, then opens out increasingly into natural history. We start with an explanation of why the ants have been so amazingly successful. The reason, we argue, is the swiftly applied and overwhelming power arising from the cooperation of colony members. Combined action at this level of efficiency is made possible by the advanced development of chemical communication: the release of a medley of substances from different parts of the body that are tested and smelled by nestmates and evoke in them, according to the substances released and the circumstance of the moment, respectively alarm, attraction, nursing, food offering, and a diversity of other activities. Ants, like humans, to put it in a nutshell, succeed because they talk so well.

The colony is the unit of meaning in the lives of ants. The workers' loyalty to it is nearly total. Perhaps as a result, organized conflict among colonies of the same species is far more frequent than human war. According to species, ants employ propaganda, deception, skilled surveillance, and mass assaults singly or in combination to overcome their enemies. In extreme cases, some fight by dropping stones on their adversaries, while others conduct slave raids to increase the size of their labor and fighting forces. But not all is harmony inside the warrior states, even those engaged in desperate territorial defense. Selfish behavior is common, especially during conflict over reproductive rights. Workers possessing ovaries sometimes compete with the queen by inserting eggs of their own into the communal nurseries. In the absence of the queen, and sometimes even in her presence, they fight for dominance. The ant colony, entomologists have discovered, is maintained by a Darwinian balance between survival by allegiance to the colony on the one side and the struggle for control within it on the other. The organization of the colony members is consequently complex and tight enough to create the equivalent of a giant, well-coordinated organism, the famous insect "superorganism."

The ants, we will show, arose amidst the dinosaurs about a hundred million years ago and spread quickly around the world. Like most such highly dominant life forms (humanity is a conspicuous exception) they have multiplied everywhere to create a plethora of species. The total number of kinds of ants at the present time probably numbers in the tens of thousands. In their expansion they have undergone a spectacular radiation of adaptive forms. This second aspect of their evolutionary achievement is the subject of the second half of the book. There we offer a tour of the whole range of formicid biodiversity, from social parasites to army ants, nomadic herders, camouflaged huntresses, and builders of temperature-controlled skyscrapers.

In our combined careers we have devoted more than eighty years of study to the ants, and we have many stories to tell, in the form of both personal anecdotes and accounts of natural history. We have also drawn heavily on the research of hundreds of other entomologists. We wish to share some of the excitement and pleasure that we and these other scientists have experienced. We hope our presentation will lead the reader to regard these insects as important in many ways to human existence.

Bert Hölldobler
Edward O. Wilson
January 3, 1994

JOURNEY
TO THE
ANTS

The Dominance of Ants

OUR PASSION IS ANTS, and our scientific discipline is myrmecology. Like all myrmecologists—there are no more than five hundred in the world—we are prone to view the Earth's surface idiosyncratically, as a network of ant colonies. We carry a global map of these relentless little insects in our heads. Everywhere we go their ubiquity and predictable natures make us feel at home, for we have learned to read part of their language and we understand certain designs of their social organization better than anyone understands the behavior of our fellow humans.

We admire these insects for their independent existence. Ants carry on in the midst of the shifting wreckage created by people, seeming not to care whether humans are present or not, so long as a little piece of minimally disturbed environment is left for them to build a nest, to search for food, and thereby to multiply their kind. City parks in Aden and San José, the steps of a Mayan temple at Uxmal, and a gutter in the streets of San Juan are among our research sites of past years, where on hands and knees we watched these minute creatures, unaware of our presence but the objects of our own lifelong curiosity and esthetic pleasure.

The abundance of ants is legendary. A worker is less than one-millionth the size of a human being, yet ants taken collectively rival people as dominant organisms on the land. Lean against a tree almost anywhere, and the first creature that crawls on you will probably be an ant. Stroll down a suburban sidewalk with your eyes fixed on the ground, counting the different kinds of animals you see. The ants will win hands down—more precisely, fore tarsi down. The British entomologist C. B. Williams once calculated that the number of insects alive on earth at a given moment is one million trillion (10^{18}). If, to take a conservative figure, 1 percent of this host is ants, their total population is ten thousand trillion. Individual workers weigh on average between 1 to 5 milligrams, according to the species. When combined, all ants in the world taken together weigh about as much as all human beings. But being so finely divided into tiny individuals, this biomass saturates the terrestrial environment.

Thus in ways that become wholly apparent only when one's line

of sight is dropped to a millimeter of the ground surface, ants lie heavily upon the rest of the fauna and flora. They envelop the lives and direct the evolution of countless other kinds of plants and animals. Ant workers are the chief predators of insects and spiders. They form the cemetery squads of creatures their own size, collecting over 90 percent of the dead bodies as fodder to carry back to their nests. By transporting seeds for food and discarding some of them uneaten in and around the nests, they are responsible for the dispersal of large numbers of plant species. They move more soil than earthworms, and in the process circulate vast quantities of nutrients vital to the health of the land ecosystems.

By specialization in anatomy and behavior, ants fill diverse niches throughout the land environment. In the forests of Central and South America, spiny red leafcutters raise fungi on pieces of fresh leaves and flowers carried into underground chambers; tiny *Acanthognathus* snare springtails with their long traplike mandibles; blind, tube-shaped *Prionopelta* squirm deep into the crevices of decaying logs to hunt silverfish; army ants advance in teeming fan-shaped formations to sweep up almost all forms of animal life; and so on through nearly endless variations among the species in the pursuit of prey, corpses, nectar, and vegetable matter. Ants reach as far as insects can in the terrestrial environment. At one extreme, species adapted for life in the deep soil almost never come to the surface. High above them, large-eyed ants occupy the forest canopies, a few kinds living in delicate nests made of leaves bound with silk.

The dominance of ants has struck us in a particularly vivid manner during visits to Finland. In cool forests stretching north to beyond the Arctic Circle, we found that these insects still dominate the land surface. In mid-May on the southern coast, with the leaves of most of the deciduous trees still only partially emerged, the sky overcast, a light rain falling, and the temperature not rising above 12°C (54° Fahrenheit, uncomfortable for scantily clothed naturalists at least), ants were active everywhere. They teemed along forest trails, atop huge moss-covered boulders, and in the grassy tussocks of bogs. In a few square kilometers could be found 17 species, one-third the known fauna of Finland.

Mound-building Formicas, red and black ants the size of houseflies,

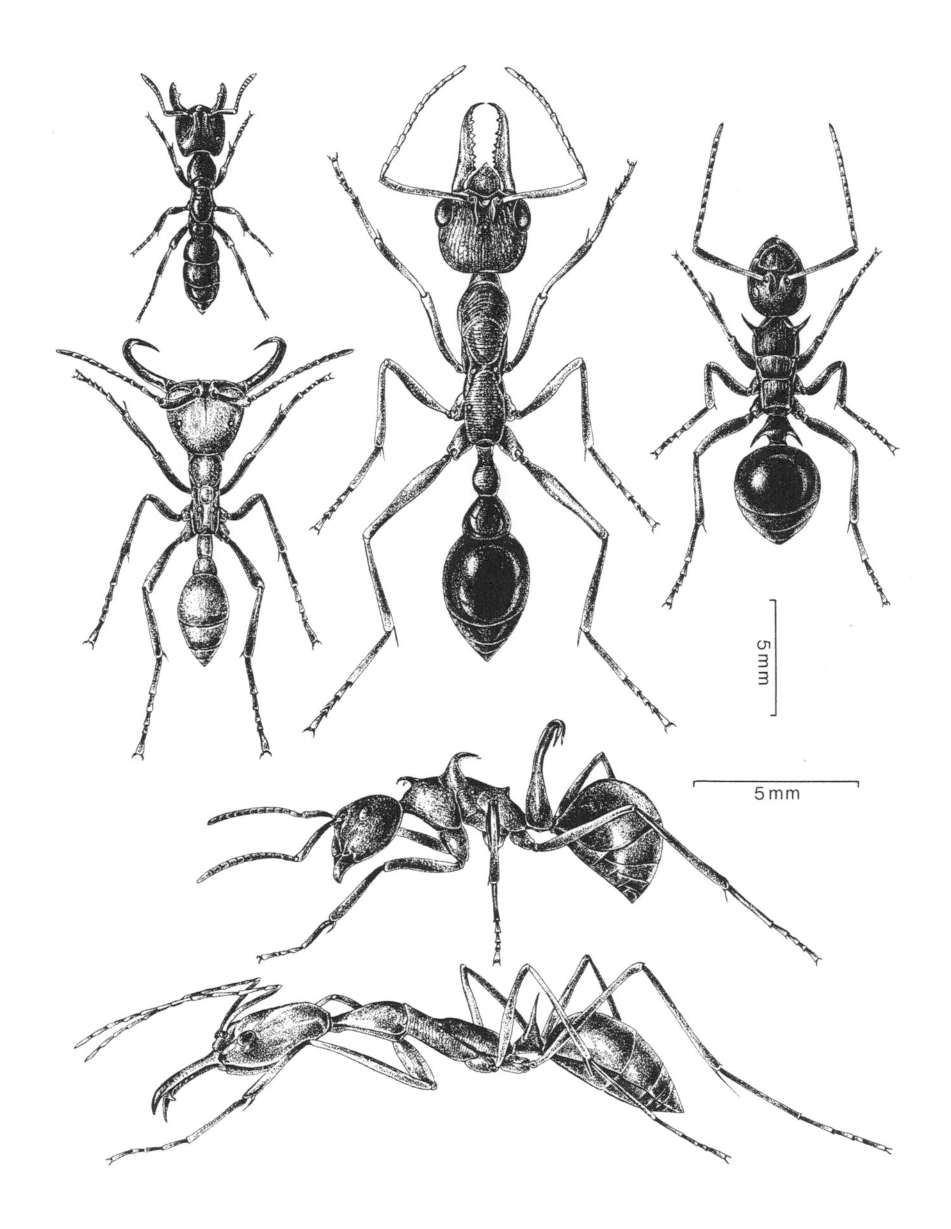

The extreme diversity of the 9,500 ant species of the world illustrated by worker ants. In the top center is a bulldog ant of the genus *Myrmecia;* to its left are a thick-bodied *Amblyopone* and a sickle-mandibled army ant of the genus *Eciton.* To the right of the bulldog ant is a multiple-spined *Polyrhachis,* and below it is another *Polyrhachis* and a long-mandibled *Odontomachus.* (Drawings by Turid Forsyth.)

A diversity of ants from South America. On the left is a long-necked *Dolichoderus;* on the right is a *Daceton,* with spines and long trap jaws. The center ants are *Pseudomyrmex* at the top and a flat turtle ant, *Zacryptocerus,* below. (Drawings by Turid Forsyth.)

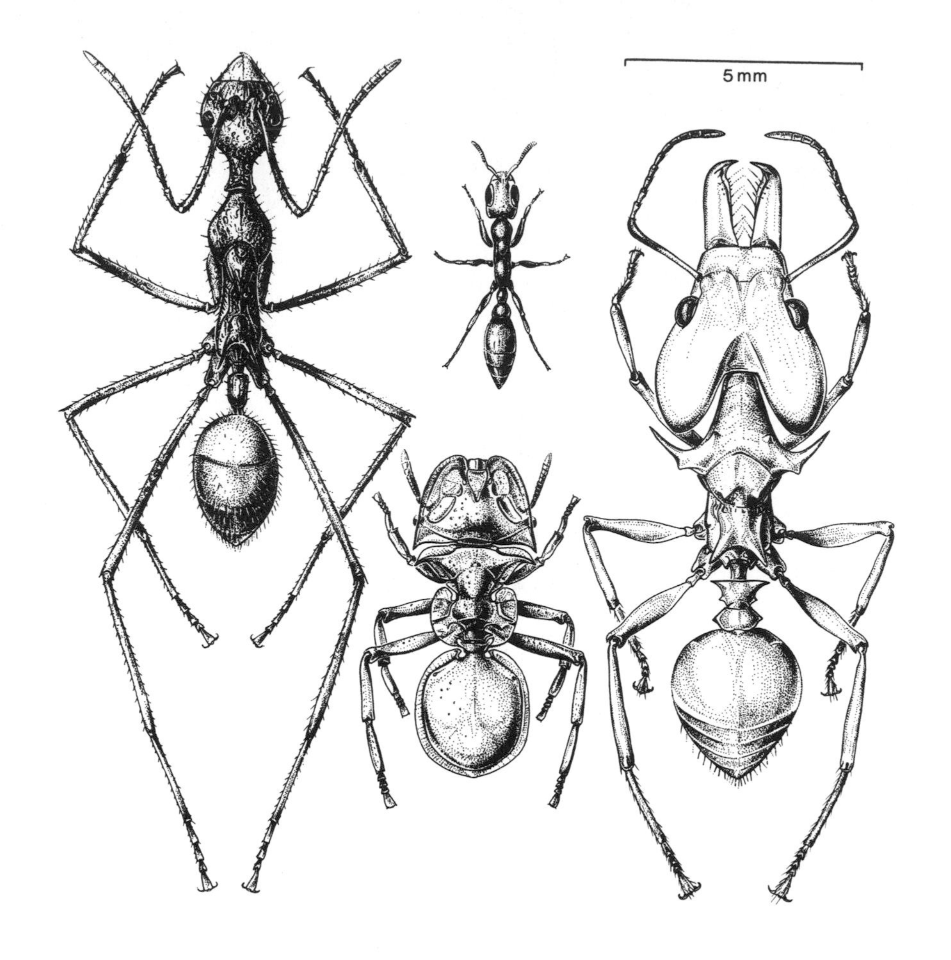

Facing page
The diversity of ants illustrated by close portraits of their heads. Clockwise from the upper left: *Orectognathus versicolor* from Australia; *Camponotus gigas* from Borneo, one of the world's largest ants; a *Zacryptocerus* from South America; and *Gigantiops destructor* from South America. (Scanning electron micrographs by Ed Seling.)

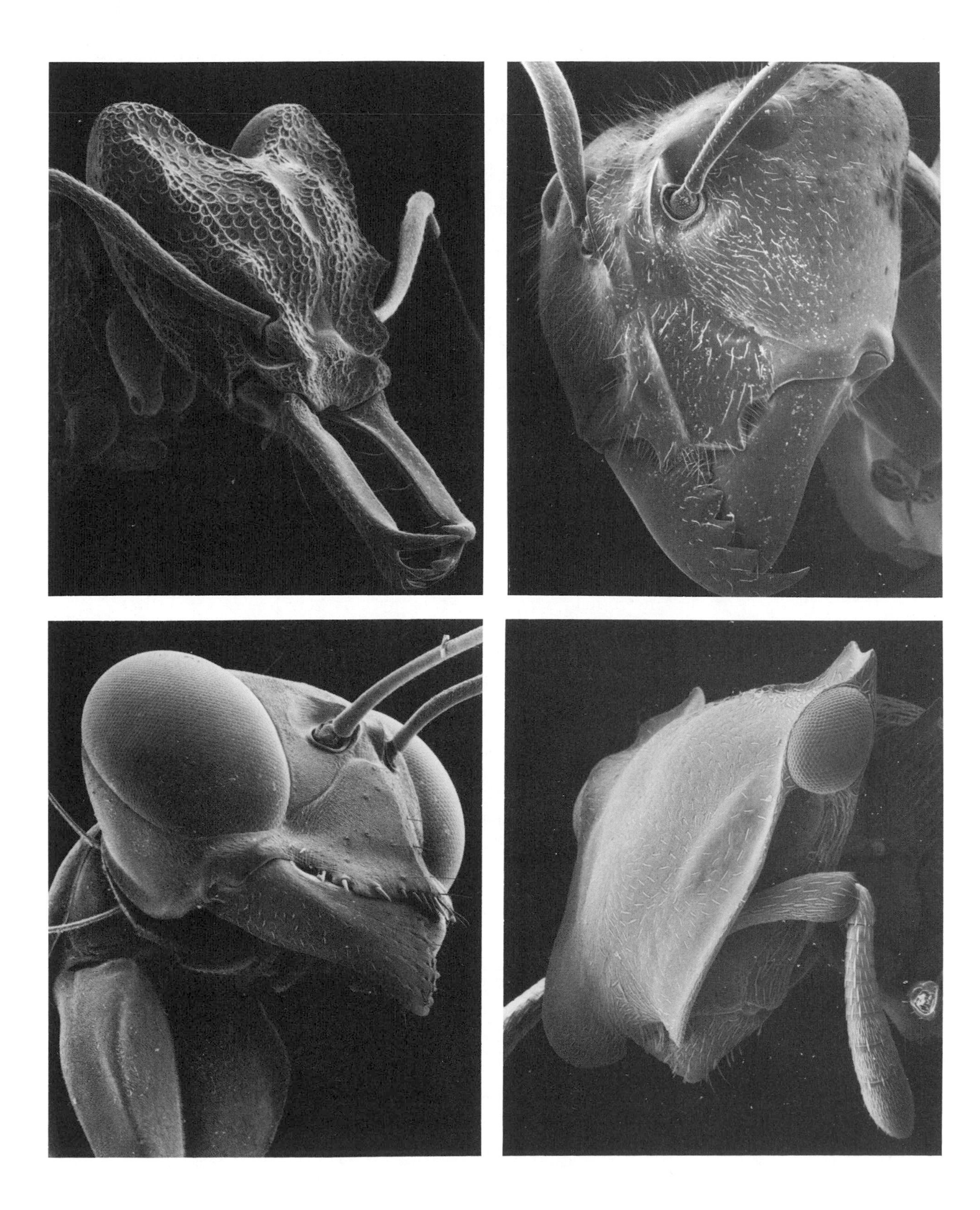

dominated the ground surface. The cone-shaped nests of several species, covered with freshly excavated soil and fragments of leaves and twigs, each housing hundreds of thousands of workers, towered a meter or more high, the equivalent for an ant of a forty-floor skyscraper. Ants seethed over the mound surfaces. They marched in columns several tens of meters long between adjacent mounds belonging to the same colony, their disciplined legions resembling heavy traffic on an intercity freeway as seen from a low-flying airplane. Other columns streamed up the trunks of nearby pine trees, where the ants attended groups of aphids and collected their sugary excrement. A small army of foragers deployed across the intervening terrain in search of prey. Some could be seen returning with caterpillars and other insects. Others were attacking colonies of smaller ants—in victory they carried the corpses of the defenders home for food.

In the forests of Finland ants are the premier predators, scavengers, and turners of soil. As we searched in company with Finnish entomologists under rocks, through the upper layers of humus, and in rotting pieces of wood strewn over the forest floor, we seldom found a patch of more than a few square meters anywhere free of these insects. Exact censuses remain to be made, but it seems likely that ants make up 10 percent or more of the animal biomass of that region.

An equal or even greater mass of living ants is found in tropical habitats. In the rain forest near Manaus, Brazil's principal city of the central Amazon, the German ecologists L. Beck, E. J. Fittkau, and H. Klinge found that ants and termites together compose nearly a third of the animal biomass: when all kinds of animals, large and small, from jaguars and monkeys down to roundworms and mites, are weighed, nearly a third of the weight consists of the flesh of ants and termites. These insects, along with the other two prevailing colonial forms, the stingless bees and polybiine wasps, make up an astonishing 80 percent of the insect biomass. And ants absolutely dominate in the canopies of the South American rain forests. In the high canopy of Peru they compose as much as 70 percent of the individual insects.

The diversity of ants in tropical localities is far higher than in Finland and other cold temperate countries. From one site of 8 hectares (20

In the Brazilian Amazon rain forest, the dry weight of all the ants is approximately four times that of all the land vertebrates (mammals, birds, reptiles, and amphibians) combined. The difference is represented here by the relative size of an ant *(Gnamptogenys)* and a jaguar. (Drawing by Katherine Brown-Wing.)

acres) in the Peruvian rain forest, we and other researchers have identified over 300 species. In a nearby locality we identified 43 species from a single *tree,* almost as many as occur in all of Finland, or all of the British Isles.

Although few such estimates of abundance and diversity have been attempted elsewhere, our strong impression is that ants and other social insects dominate terrestrial habitats in like degree throughout most of the rest of the world. All together, these creatures seem likely to constitute half or more of the insect biomass. Consider the following disproportion: only 13,500 species of highly social insects are known (9,500 of which are ants) out of a grand total of 750,000 insect species that have been recognized to date by biologists. Thus more than half the living tissue of insects is made up of just 2 percent of the species, the fraction that live in well-organized colonies.

We believe that the anomaly is due in large part to a struggle for existence based on harsh, direct competitive exclusion. The highly social insects, particularly the ants and termites, occupy center stage in the terrestrial environment, in the literal sense of having evicted silverfish, hunting wasps, cockroaches, aphids, hemipteran bugs, and most other kinds of solitary insects from the most desirable, stable nest sites. The solitary forms tend to occupy more remote and transient resting places, such as distant twigs, the extremely moist or dry or excessively crumbling pieces of wood, the surfaces of leaves, and the newly exposed soil of stream banks. As a rule, they are also either very small, or fast moving, or cleverly camouflaged, or heavily armored. At the risk of oversimplification, we envisage an overall pattern of ants and termites at the ecological center, solitary insects at the periphery.

How have ants and other social insects come to lord over the terrestrial environment? In our opinion their edge comes directly from their social nature. There is strength in numbers, if all the minions are programmed to act in concert. This quality is not, of course, unique to the insects. Social organization has been one of the most consistently successful strategies in all of evolutionary history. Consider that the coral reefs, which cover much of the floor of the shallow tropical seas, are composed of colonial organisms, sheetlike masses of anthozoan zooids, to be exact, which are distant relatives of the solitary and less abundant jellyfish. And that human beings, the most dominant mammals in geological history, are also by far the most social.

The most advanced social insects, those forming the biggest and most complicated societies, have attained this rank through a combination of three biological traits: the adults care for the young; two or more generations of adults live together in the same nest; and the members of each colony are divided into a reproductive "royal" caste and a nonreproductive "worker" caste. This elite group, which entomologists call eusocial (meaning "truly" social), is made up mainly of four familiar groups:

All of the *ants,* composing in formal taxonomic classification the family Formicidae of the order Hymenoptera, contain about 9,500 species known to science and at least twice that number of species remaining to be discovered, most of which are confined to the tropics.

Some of the *bees* are eusocial. At least ten independent evolutionary lines within the Halictidae (sweat bees) and Apidae (honeybees, bumblebees, and stingless bees) have reached the eusocial level. They contain about a thousand species known to science. A much larger number of bee species are solitary, including a large majority of the sweat bees.

Some of the *wasps* are also eusocial. About 800 species in the family Vespidae and a handful in the Sphecidae are known to have reached this evolutionary level. But they represent a minority, just as in the bees. Tens of thousands of other wasp species, scattered through many taxonomic families, are solitary.

All *termites*, composing an entire order on their own (the Isoptera), are eusocial. Descended from cockroach-like ancestors as far back as 150 million years ago, early in the Mesozoic Era, these curious insects have converged in evolution toward ants in superficial appearance and social behavior, but they have nothing else in common. About 2,000 species of termites are known to science.

In our view, the competitive edge that led to the rise of the ants as a world-dominant group is their highly developed, self-sacrificial colonial existence. It would appear that socialism really works under some circumstances. Karl Marx just had the wrong species.

The advantage of ants comes to the fore in the arena of labor efficiency. Consider the following scenario. A hundred solitary female wasps are pitted against an ant colony with the same number of workers, also all female. The two aggregations nest side by side. In a typical daily action, one of the wasps digs a nest and captures a caterpillar, a grasshopper, a fly or some other prey to serve as provender for her offspring. Next she lays an egg on the prey and closes the nest. The egg will hatch into a grublike larva, which will feed on the insect provided and in time emerge as a new adult wasp. If the mother wasp falters in any one of the serial tasks up to the sealing of the nest, or if she tries to perform them in the wrong sequence, the entire operation fails.

Nearby an ant colony, functioning as a *social unit*, overcomes all these difficulties automatically. A worker starts to dig a chamber to expand the colonial nest, where larvae will eventually be moved and fed to produce additional members of the colony. If the ant fails at any step of her sequence, all the necessary tasks will probably be finished anyway, so that the colony will continue to grow. A sister worker will simply move in and complete the excavation; other sisters can be counted on to transport larvae to the chamber, and still others to bring food. Many of the ants are "patrollers." On stand-by status, these individuals travel restlessly through the corridors and rooms, addressing each contingency they encounter, switching back and forth from one task to the other as needed. They complete the sequence of steps more reliably and finish in less time than could solitary laborers. They are like gangs of factory workers who move back and forth among the assembly lines according to momentary need and opportunity, improving the efficiency of the overall operation.

The grand strategy of social life becomes most obvious during territorial disputes and competition for food. Ant workers enter combat more recklessly than do solitary wasps. They can act like six-legged kamikazes. The solitary wasp has no such choice. If she is killed or injured, the Darwinian game is over, just as it would be if she had blundered during her labors and aborted the necessary rounds of nest construction and provisioning. Not so the ant. She is nonreproductive to start with and if lost will be quickly replaced by a new sister born back in the nest. So long as the mother ant queen is protected and continues to lay eggs, the death of one or a few workers will have little effect on the representation of the colony members in the future gene pool. What counts is not the total population of the colony but the number of virgin queens and males released into the nuptial flights that are successful at starting new colonies. Suppose that the war of attrition between ants and solitary wasps continues until almost all the ant workers are destroyed. So long as the queen lives through the encounter, the ant colony wins. The queen and surviving workers will rebuild the worker population rapidly, allowing the colony to reproduce itself by producing virgin queens and males. The solitary wasp, the equivalent of an entire colony, will long since have perished.

This built-in competitive superiority of colonies against wasps and other solitary insects means that colonies can retain prime nest sites and feeding areas for the natural life of the mother queen. In some species she lives more than twenty years. In others, where young queens return home after being mated, the colony has even greater potential: the nests and territories can be passed from one generation to the next. To heredity, then, is added the inheritance of property. The nests of mound-building ants, such as the Formica wood ants of Europe, often last for many decades, churning out queens and males year after year. Such colonies are in fact potentially immortal, even though the individual queens at their center are continually dying and being replaced.

There is still more to the power of the superorganism that is an ant colony. Building larger nests than individual solitary wasps and holding on to them for longer periods of time, ant colonies devise physical structures elaborate enough to regulate the climate. Workers of some species drive tunnels deep beneath the surface to reach soil containing more moisture. Those of others excavate galleries and chambers that radiate outward in a way that increases the flow of fresh air through their living quarters. During short-term emergencies architecture is augmented by rapid mass responses. In many species, when the nest dries out during droughts or excessive heat, workers form loosely organized bucket brigades, dashing back and forth in short lines, passing water from mouth to mouth, and finally regurgitating it onto the nest floor and walls. When enemies break through the nest wall some of the workers attack the invaders while others rescue the young or rush to repair the damage.

Colonial life may be an ancient phenomenon by human standards, but it is a relatively recent development in the overall evolution of the insects. It covers only about half of their geological time on earth. Insects were among the first creatures to colonize the land, arising as far back as the Devonian period, some 400 million years ago. They diversified richly in the swamps of the Coal Age that followed. By Permian times, about 250 million years ago, forests teemed with cockroaches, hemipteran bugs, beetles, and dragonflies not much different from those living today, mingled with beetle-like protelytropterans; protodonates, resembling huge dragonflies, with wings up to 3 feet across; and other insect orders now extinct. The first termites probably arose in the Jurassic or early Cretaceous Periods, roughly 200 million years before the present, and ants, social bees, and social wasps in the Cretaceous Period some 100 million years later. The eusocial insects as a whole, particularly the ants and termites, became dominant among the insects no later than the beginning of the Tertiary Period, 50 to 60 million years ago.

The sheer magnitude of this history, stretching back more than a hundred times beyond the entire life span of the genus *Homo*, presents a paradox. Why, if colonial life has such great advantages for insects, was it delayed for 200 million years? And why, 200 million years after the innovation finally occurred, aren't all insects colonial? These questions are better asked by turning them around: What advantages, not yet mentioned, might solitary life have over social life? The answer, we believe, is that solitary insects breed faster and do better with limited, ephemeral resources. By picking up the pieces left over by the ants and other eusocial insects, they fill transient niches.

It may seem odd to say that the highly social insects breed more slowly than their solitary counterparts. Colonies are after all little factories crowded with workers devoted to the mass production of new nestmates. But the fact of central importance is that the colony, not the workers, is the unit of reproduction. Where every solitary wasp is a potential mother or father, only one out of hundreds or thousands of the members of an ant colony can fill that role. In order to create virgin queens capable of founding new colonies, the mother colony—the superorganism, the unit of reproduction—must first produce a crop of workers. Only then can it reach the stage equivalent to sexual maturity in a solitary organism.

Being a massive organism, the colony must also have a large base from which to operate. It dominates the logs and fallen branches but concedes the scattered leaves and flakes of bark to the fast-moving, fast-breeding solitary insects. It controls the stable river banks but gives up the transient mud bars farther out. It spreads more slowly from one feeding place to the next because all the population must be mobilized before any one member can safely emigrate.

Solitary insects are therefore better pioneers. They can reach small windfalls in distant places—a seedling in a patch of new ground, a twig washed downstream, a new sprig of leaves—more quickly and flourish there for longer periods of time. Ant colonies in contrast are ecological juggernauts. They take time to grow up, they move about slowly, but once in motion they are very difficult to stop.

For the Love of Ants

DURING the 1960s and 1970s the scientific study of ants accelerated, swept forward by the general revolution in biology. In short order entomologists discovered that colony members communicate most of the time through the taste and smell of chemicals secreted from special glands throughout the body. They conceived the idea that altruism evolves by kin selection, the Darwinian advantage gained by the selfless care of brothers and sisters, who share the same altruistic genes and thus transmit them to future generations. And they established that the elaborate caste systems—queens, soldiers, workers, the signature trait of the ant societies—are determined by food and other environmental factors and not by genes.

In the fall of 1969, in the midst of this exciting period, Hölldobler knocked on Wilson's office door at Harvard University at the beginning of a term as Visiting Scholar. Although we didn't think of ourselves that way at the time, we met as representatives of two scientific disciplines, born of different national scientific cultures, whose synthesis was soon to lead to a better understanding of ant colonies and other complex animal societies. One discipline was ethology, the study of behavior under natural conditions. This branch of behavioral biology, conceived and developed mostly in Europe during the 1940s and 1950s, differed sharply from traditional American psychology by its emphasis on the importance of instinct. It also stressed how behavior adapts animals to those special parts of the environment on which the survival of the species depends. It singled out which enemies to avoid, which food items to hunt, the best places to build nests, where and with whom and how to mate, and so on through each step of the intricate life cycle. Ethologists were above all (and many so remain) naturalists of the old school, outfitted with muddy boots, waterproof notebooks, and sweat-soaked binocular straps chafing the neck. But they were also modern biologists who used experiments to dissect the elements of instinctive behavior. In combining these two approaches to become more scientific, they discovered "sign stimuli," the relatively simple cues that trigger and guide stereotyped behaviors in animals. For example, a red belly on

a male stickleback fish, really no more than a red spot to the animal eye, provokes a full territorial display in a rival stickleback male. The males are programmed to react to the splash of color and not to the look of a whole fish, at least not to what we as human beings see in a whole fish.

The annals of biology are now filled with such examples of sign stimuli. The smell of lactic acid guides the yellow-fever mosquito to its victim; a flash of ultraviolet reflecting wings identifies a male sulfur butterfly to the waiting female; a dash of glutathione in the water causes the hydra to stretch its tentacles in the direction of suspected prey; and so on bit by bit through the vast repertory of animal behavior, now well understood by ethologists. Animals, they realized, survive by responding swiftly and precisely to the fast-moving environment, hence the reliance on simple pieces of their sensory world. The responses in turn must often be complex, unlike the sign stimuli, and delivered in exactly the right manner. Animals are rarely given a second chance. And because all this repertory has to be accomplished with little or no opportunity to learn anything in advance, it must have a strong automatic, genetic basis. The nervous system of animals, in short, must to a substantial degree be hard-wired. If that much is true, the ethologists reasoned, if behavior is hereditary and shaped in a manner peculiar to each species, then it can be studied element by element, with the time-honored techniques of experimental biology, as though it were a piece of anatomy or a physiological process.

By 1969 the idea that behavior could be broken apart into atomic units had energized the entire generation of behavioral biologists to which we belonged. The effect was enhanced personally for us by the fact that one of the founders of ethology was a great Austrian zoologist who was a professor at the University of Munich in Germany with interests similar to our own. Karl von Frisch was and remains one of the most famous biologists in the world, praised for his discovery of the waggle dance, the elaborate movements in the hive by which honeybees inform their nestmates about the location and distance of food finds outside. The waggle dance remains to this day the closest approach to a symbolic language known in the animal kingdom. More generally, von Frisch was esteemed among biologists for the ingenuity and elegance of his many experiments on animal senses and behavior. In 1973 he shared the Nobel prize in Physiology or Medicine with his fellow Austrian Konrad Lorenz, former director of the Max Planck Institute for Behavioral Physiology in Germany, and with Nikko Tinbergen of the Netherlands, a professor at Oxford University in England, for the leading role the three men played in the development of ethology.

The second watershed tradition leading to a new understanding of animal societies was largely of American and British origin, with approaches radically different from those of ethology. It was population biology, the study of the properties of entire populations of organisms, how they grow as an aggregate, spread across the landscape, and, inevitably, retreat and vanish. The discipline relies as much on mathematical models as on field and laboratory studies of live organisms. Very like demography, it deduces the fate of populations by tracing the birth, death, and movements of the individual organisms, in order to plot overall trends. It also tracks gender, age, and the genetic makeup of the organisms.

As we began our collaboration at Harvard, we understood that ethology and population biology fit together wonderfully well in the study of ants and other social insects. Insect colonies are little populations. They can be understood best by following the life and death of the swarming legions that compose them. Their hereditary makeup, especially the kinship of their members, predetermines their cooperative nature. The things we learn from ethology about the details of communication, colony founding, and caste come together and make complete sense only when they are viewed as evolutionary products of whole colony populations. That in a nutshell is the basis for the new discipline of sociobiology, the systematic study of the biological basis of social behavior and of the organization of complex societies.

As we began conversations on this synthesis and our research agendas, Wilson was a professor at Harvard, 40 years old; Hölldobler, at 33, was on leave from a lectureship at the University of Frankfurt. Three years later, after a brief return to teach at Frankfurt, Hölldobler was invited to Harvard as a full professor. Thereafter the friends shared the fourth floor of the newly constructed laboratory wing of the university's Museum of

Comparative Zoology until, in 1989, Hölldobler returned to Germany to direct a department entirely devoted to the study of social insects in the newly founded Theodor Boveri Institute of Biosciences of the University of Würzburg.

Science is said to be the one culture that truly rises above national differences, melding idiosyncratic differences into a single body of knowledge that can be simply and elegantly expressed and generally accepted as true. We entered its domain by markedly different routes of academic tradition, but impelled by a common childhood pleasure in the study of insects and by the approval and encouragement of adults at a critical time of our mental development. To put the matter as simply as possible we, having entered our bug period as children, were blessed by never being required to abandon it.

For Bert, the beginning was on a lovely early summer day in Bavaria just before massive air raids brought World War II home to Germany. He was 7 years old and had just been reunited with his father, Karl, a doctor on duty with the German army in Finland. The elder Hölldobler had obtained a furlough to visit his family at Ochsenfurt. He took Bert on a walk through the woods, just to look around and talk. But this was not quite an ordinary stroll. Karl, an ardent zoologist, had a particular interest in ant societies. He was an internationally known expert on the many curious small wasps and beetles that live in ant nests. It was natural on this occasion for him to turn over rocks and small logs along the trail to see what was living underneath. Rooting through the soil to see its teeming life is, he understood well, one of the pleasures of entomology.

One rock sheltered a colony of large carpenter ants. Caught for an instant in the sunlight, the shiny blackish-brown workers rushed frantically to seize and carry grublike larvae and cocoon-encased pupae (their immature sisters) down the subterranean channels of the nest. This sudden apparition riveted young Bert. What an exotic and beautiful world, how complete and well formed! A whole society had revealed itself for an instant, then trickled magically out of sight, like water into dry soil, back to the subterranean world to resume a way of life strange beyond imagination.

After the war the Hölldobler home in the little medieval town of

Ochsenfurt, close to Würzburg, was filled with pets, at various times including dogs, mice, guinea pigs, a fox, fish, a large salamander called an axolotl, a heron, and a jackdaw. A guest of special interest to Bert was a human flea, which he kept in a vial and allowed to feed on his own blood, in an early attempt at scientific research.

Above all, encouraged by the example of his father and the loving patience of his mother, Bert kept ants. He gathered live colonies and studied them in artificial nests, learning the local species, drawing their distinctive anatomical traits, and observing their behavior. All the while his enthusiasms bubbled over. On top of everything else he collected butterflies and beetles as yet another hobby. He was imprinted on the diversity of life, the die was cast, and his hopes now centered on a career in biology.

In the fall of 1956, Bert entered the nearby University of Würzburg, intending to teach biology and other sciences in high school. By the time he took his final examination, however, he had lifted his horizons. He gained admittance to the graduate program of the university, now aiming for a doctoral degree. His teacher at this new level was Karl Gösswald, a specialist on wood ants. These large red and black insects, swarming by the millions per hectare, build mound nests that dot the forests of northern Europe. Gösswald wished to develop propagation methods by which the ants could control the caterpillars and other pests of the forest vegetation, without the intervention of insecticides. For generations European entomologists had noticed that whenever an outbreak of leaf-eating insects occurred, trees around the ant mounds remained healthy, with their foliage more or less intact. The protection was clearly the result of predation of the pests by the ants. Direct counts revealed that one wood-ant colony can harvest in excess of 100,000 caterpillars in a single day.

An early pioneer of forest entomology, Karl Escherich, spoke of the "green islands" that exist under the protective shield of the wood ants. Escherich was a student at the University of Würzburg in the 1890s, working under the tutelage of Theodor Boveri, at that time the most celebrated embryologist in the world. By fortunate coincidence, William Morton Wheeler, later to become America's leading myrmecologist, was at that time also an embryologist, and he visited Würzburg for two years as a young scholar. He was soon to switch his main research activity to ants. (Later, in 1907, he settled at Harvard as professor of entomology—thus he was Wilson's predecessor.) He conveyed his early enthusiasm for ants to young Escherich, who, partly as a result of Wheeler's influence, abandoned an interest in medicine and turned to forest entomology. His multivolume masterwork on the subject, completed in later life, influenced an entire generation of German researchers, among them Karl Gösswald. Initially, however, it was none other than Karl Hölldobler, then an advanced student of medicine and zoology at Würzburg, who introduced Gösswald to myrmecology. He encouraged the younger student to explore the rich ant fauna of the limestone area along the Main River in Franconia, a part of northern Bavaria. The work became the basis of Gösswald's doctoral thesis. So the two lineages run as follows: first, Wheeler–Escherich–Karl Hölldobler–Gösswald–Bert Hölldobler and, second, Wheeler–Frank M. Carpenter (Wilson's teacher at Harvard)–Wilson, starting in Würzburg with Wheeler, then separating, and finally, as we shall see, looping back to touch the German enterprise again at Harvard. Such is the reticulate structure of heritage in the scientific world.

Bert was far from exclusively guided by Gösswald while at Würzburg. Because of his father's friendship with other myrmecologists in the postwar years, he met many fellow enthusiasts before he entered the university. Among them were Heinrich Kutter of Switzerland and Robert Stumper of Luxembourg. Bert was attracted to forest entomology, but the mental gyroscope he had acquired as a child brought him back inevitably to the ants. At that time he was also inspired by Hans-Jochem Autrum, who gave lectures in zoology and, as one of the foremost neurophysiologists in the world, served as an inspiring role model.

One of Bert's first assignments, while he was still an undergraduate student, was a trip to Finland to conduct a north-to-south survey of wood ants. It was a full-time job, but Bert could not keep his eyes off the equally prominent carpenter ants, including the species that had conjured magic beneath the stone at Ochsenfurt. He felt nostalgia in visiting the forests of Karelia, where his father had spent the war under difficult and often dangerous conditions. Now it had become the scene for a leisurely exploration of a little-known fauna. Much of Finland was, and remains, a wilderness country, especially the northern reaches. Searching through its forests and glades, filled with mostly unstudied insect life, cemented Bert's commitment to field biology.

His mind was turning away from the kind of applied entomology emphasized by Karl Gösswald, and more toward the basic research favored by his instincts and early training. About three years after the Finland trip he learned of a graduate studies program at the University of Frankfurt headed by Martin Lindauer, one of von Frisch's most gifted students, and generally regarded as the great man's intellectual successor. In the 1960s Lindauer and his own protégés were in the midst of an exciting new wave of research on honeybees and stingless bees, and Frankfurt had become the center of what is aptly called the von Frisch–Lindauer school of animal behavior studies. Its tradition was not just a professional staff and a set of techniques but a philosophy of research based on a thorough, loving interest in—a *feel* for—the organism, especially as it fits into the natural environment. Learn the species of your choice every way you can, this whole-organismic approach stipulates. Try to understand, or at the very least try to imagine, how its behavior and physiology adapt it to the real world. Then select a piece of behavior that can be separated and analyzed as though it were a bit of anatomy. Having identified a phenomenon to call your own, press the investigation in the most promising direction. And don't hesitate to ask new questions along the way.

Every successful scientist has a small number of personal ways of coaxing discoveries out of nature. Von Frisch himself had two in which he attained great mastery. The first was the close examination of the flight of honeybees from hive to flowers and back again, a part of the life of bees that can be easily watched and manipulated. The second was the method of behavioral conditioning by which von Frisch combined stimuli, such as the color of a flower or the smell of a fragrance, with a subsequent meal of sugar water. In later tests, bees and other animals will then respond to the stimuli, provided they are strong enough to be detected. Using this simple technique, von Frisch was the first to demonstrate conclusively that insects can see color. He discovered that honeybees can also see polarized light, a capacity not possessed by human beings. The bees use polarized light to estimate the position of the sun, and take a compass reading, even when the sun is hidden behind clouds.

After Hölldobler completed the requirements for his doctoral degree at Würzburg in 1965, he moved to Frankfurt to work under Lindauer. The German doctoral students and young postdoctoral researchers he joined there were an outstanding group of young scientists, destined for leadership in the study of social insects and behavioral biology. They included Eduard Linsenmair, Hubert Markl, Ulrich Maschwitz, Randolf Menzel, Werner Rathmayer, and Rüdiger Wehner. Wehner was later to move to the University of Zurich, where he pioneered in the visual physiology and orientation of bees and ants.

This circle and these environs proved to be Bert's natural intellectual home. Given freedom to study the subject that had enchanted him since childhood, and encouraged by von Frisch himself, he set to work full time on new projects in the behavior and ecology of ants. In 1969 he received his habilitation, the equivalent of a second doctorate and the certification needed in Germany to become an instructor with classes of one's own. He began his new career by visiting Harvard University for two years, then returned briefly to teach at the University of Frankfurt, and in 1972 came back to Harvard. Thus began the main part of his twenty-year collaboration with Wilson.

In 1945, not long after Hölldobler's childhood encounter with the Ochsenfurt ant colony, Ed Wilson had recently moved from his hometown of Mobile to Decatur, a northern Alabama city named for Stephen Decatur, the War of 1812 hero renowned for his postprandial toast, "Our country! May she always be right; but our country, right or wrong." True to its honorand, Decatur was a municipality of right thinking and attention to civic duty. Having reached 16 years of age, Ed, also known as Bugs or Snake to his friends, believed he should be preparing for his future in a serious manner. The time had come to say farewell to the Boy Scouts of America, where he had earned the rank of Eagle Scout, to move past mere snake catching and bird watching, to defer involvement with girls—for a while anyway—and above all to give careful thought to his future career as an entomologist.

A large mound of the red wood ant *Formica polyctena* in the primeval forest of Finland. The picture was taken during Bert Hölldobler's first exploration of the Finnish ant fauna in 1960, and shows his Finnish myrmecological friend Heikki Wuorenrinne.

Nuptial flights in the Chihuahuan desert in the southwestern United States. After heavy summer rains have softened the soil many ant species engage in mating activities. The winged males and females of *Forelius pruinosus* climb little bushes, which they use as launching pads for the nuptial flight.

Facing page
During their nuptial flights male and female ants of many species assemble in swarms to mate. As shown in the upper photograph, harvester ants *(Pogonomyrmex desertorum)* fly upwind and gather on mesquite bushes. On arrival the males release a strong-smelling scent that attracts more females and other males to the communal mating place. In the lower photograph, swarms of a species of the ant genus *Pheidole* form over the hot asphalt surface of a country road leading through the Arizona desert.

Mating frenzy in the American harvester ant *Pogonomyrmex rugosus*. Thousands of males and females gather in specific areas on the ground. In the foreground can be seen a male copulating with a young female. (Painting by John D. Dawson, courtesy of the National Geographic Society.)

Queens and males of the American harvester ant *Pogonomyrmex barbatus* also gather in specific sites to mate. The young queens are always greatly outnumbered by the males. Often 10 or more males attempt to mate with a female at the same time.

The hours after mating are the most dangerous period for the queens. As they shed their wings and search for a suitable place to dig their founding nest, a majority fall prey to other ant species, lizards, or spiders. Here a crab spider has caught a queen of *Pogonomyrmex maricopa.*

Facing page
Once the foundress queen has raised the first workers, the colony grows swiftly. A queen of *Pheidole desertorum* is shown in the upper photograph surrounded by her first workers, eggs, larvae, and pupae. The lower picture shows the first square-headed soldiers and a few newly eclosed, still lightly colored workers.

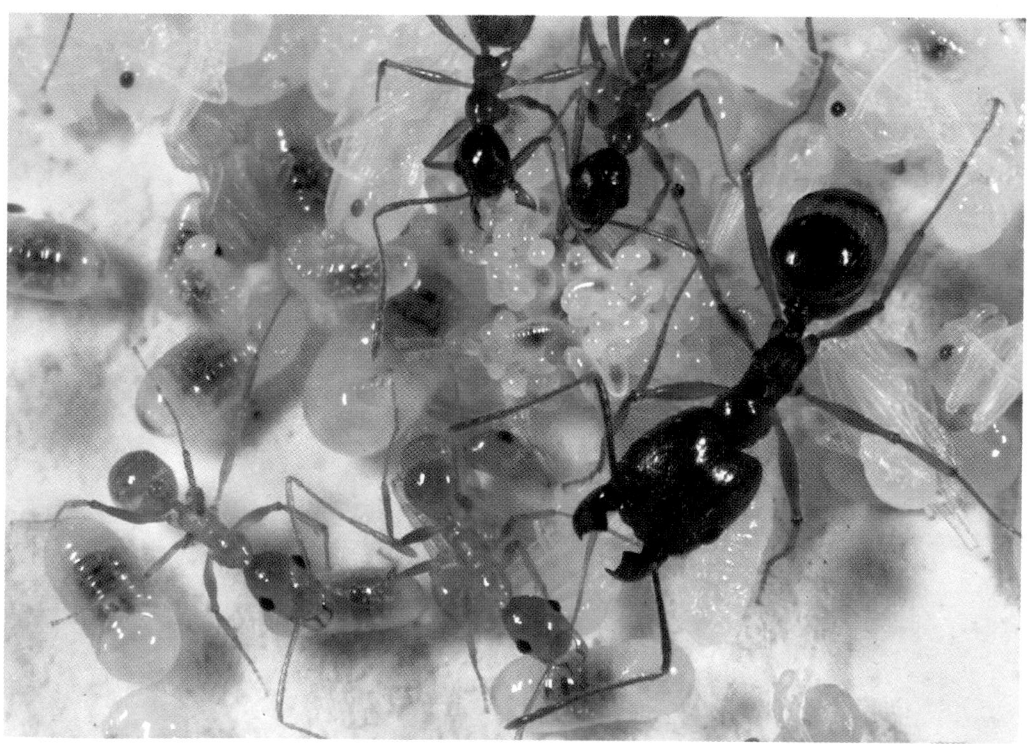

Camponotus perthiana. One queen of this Australian species lived more than 23 years in a laboratory nest. During this time she produced hundreds of worker offspring.

He believed that the best route was to acquire expertise on some group of insects that offered opportunities for scientific discovery. His first choice was the Diptera, the order of flies, and especially the family Dolichopodidae, sometimes called long-legged flies, glittering little metallic green and blue insects found dancing in mating rituals on the sunlit upper surfaces of leaves. The opportunities were extensive; over a thousand kinds occur in the United States alone, and Alabama itself was mostly unexplored. But Ed was thwarted from fulfilling this first ambition. The war had shut off the supply of insect pins, the standard equipment used to preserve and store specimens of flies. These special black, ball-headed needles were manufactured in Czechoslovakia, which at that time was still under German occupation.

He needed a kind of insect that could be preserved with equipment immediately at hand. So he turned to ants. The hunting grounds were the wooded lots and fields along the Tennessee River. The equipment, consisting of 5-dram prescription bottles, rubbing alcohol, and forceps, could be bought in small-town pharmacies. The text was William Morton Wheeler's 1910 classic *Ants,* which he bought with earnings from his morning delivery route for the city newspaper, the *Decatur Daily.*

Six years previously the seeds had been set for a career as a naturalist, but not in the outdoors of Alabama. At that time Ed's family lived in the heart of Washington, D.C., close enough for a short automobile drive to the Mall for Sunday outings and, more important for an embryo naturalist, within walking distance of the National Zoo and Rock Creek Park. Adults saw this part of the capital for what it was in human terms, a decaying urban neighborhood close to the high-energy center of government. For a 10-year-old, however, it was a region teeming with the fragments and emissaries of an enchanted natural world. On sunny days, carrying a butterfly net and cyanide killing jar, Ed wandered through the zoo to stand as close as possible to elephants, crocodiles, cobras, tigers, and rhinoceroses, and then, a few minutes later, he walked onto the back roads and woodland paths of the park to hunt for butterflies. Rock Creek Park was the Amazon jungle writ small, in which Ed, often accompanied by his best friend, Ellis MacLeod (now a professor of entomology at the University of Illinois), could live in his imagination as an apprentice explorer.

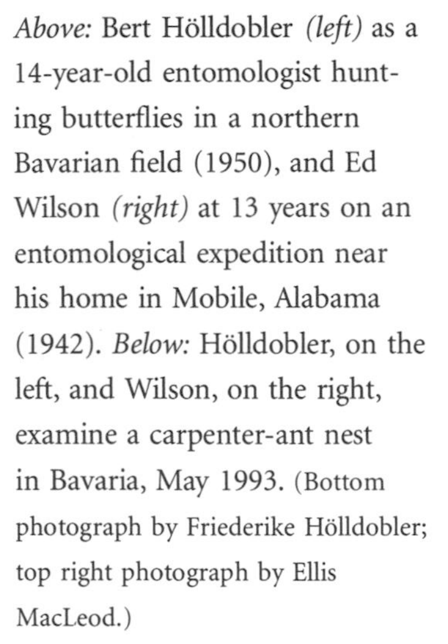

Above: Bert Hölldobler *(left)* as a 14-year-old entomologist hunting butterflies in a northern Bavarian field (1950), and Ed Wilson *(right)* at 13 years on an entomological expedition near his home in Mobile, Alabama (1942). *Below:* Hölldobler, on the left, and Wilson, on the right, examine a carpenter-ant nest in Bavaria, May 1993. (Bottom photograph by Friederike Hölldobler; top right photograph by Ellis MacLeod.)

On other days Ellis and Ed took streetcar rides to the National Museum of Natural History to explore the exhibitions of animals and habitats and pull out trays of pinned butterflies and other insects from around the world. The diversity of life displayed in this great institution was dazzling and awe-inspiring. The curators of the National Museum seemed knights of a noble order, educated to unimaginably high levels. The director of the National Zoo loomed even more heroic in this 1939 city of civic opportunity. He was William M. Mann, by odd coincidence a myrmecologist himself, a former student of William Morton Wheeler at Harvard who had studied ants at the National Museum and then transferred to the National Zoo as its director.

In 1934 Mann had published an article on his original scholarly interest, "Stalking Ants, Savage and Civilized," in the *National Geographic.* Ed eagerly studied the piece and then went out to search for some of the species in Rock Creek Park, excited by the knowledge that the author himself worked close by. One day he had an experience similar to Bert's epiphanous encounter with the carpenter-ant colony of Ochsenfurt. Climbing up a wooded hillside with Ellis MacLeod, he pulled away the bark of a rotting tree stump just to see what lived underneath. Out poured a roiling mass of brilliant yellow ants, emitting a strong lemony odor. The chemical substance, as later research was to reveal (by Ed himself, in 1969), was citronellal, and the worker ants were expelling it from glands in their heads to warn nestmates and drive off enemies. The ants were "citronella ants," members of the genus *Acanthomyops,* whose workers are nearly blind and completely subterranean. The force in the stump quickly thinned and vanished into the dark interior. But it left a vivid, lasting impression on the boy. What netherworld had been briefly glimpsed?

In the fall of 1946 Ed arrived at the University of Alabama, at Tuscaloosa. Within days he called on the chairman of the Biology Department, preserved ant collection in hand, thinking it normal or at least not outrageous for a beginning student to announce his professional plans in such a manner and to begin, as part of undergraduate studies, research in the field of his choice. The chairman and other biology professors did not laugh or wave him away. They were gracious to the 17-year-old. They gave him laboratory space, a microscope, and frequent warm encouragement. They took him along on field trips to natural habitats around Tuscaloosa and listened patiently to his explanations of ant behavior. This relaxed, supportive ambience was formative in a decisive manner. Had Ed gone to Harvard, where he now teaches, and been thrust into a packed assemblage of valedictorian overachievers, the results might have been different. (But perhaps not. There are many odd niches at Harvard where eccentrics flourish.)

In 1950, with bachelor's and master's degrees completed, Ed moved to the University of Tennessee to begin work on his Ph.D. There he might have remained, since the southern states and their rich ant faunas seemed world enough. But he had fallen under the spell of a distant mentor, William L. Brown, seven years his senior, who was just then completing a doctorate at Harvard. Uncle Bill, as he was to be affectionately called by fellow myrmecologists in later years, was a soulmate, fixated on the subject of ants. Brown took a global approach to these insects, thinking the faunas of all countries equally interesting. His spirit was deeply professional and responsible, seeking legitimacy for small creatures all too easily waved aside. Our generation, he explained to Ed, must revamp biological knowledge and reclassification of these wondrous insects, and assign them major scientific importance in their own right. And don't be intimidated, he added, by the achievements of Wheeler and other famous entomologists of the past. These people are overrated to a stultifying degree. We can and will do better; we must. Take pride, be careful in mounting your specimens, obtain reprints for ready reference, widen your studies to many kinds of ants, expand your interests beyond the southern United States. And while you are at it, find out what dacetine ants eat (Ed then confirmed that dacetines prey on springtails and other soft-bodied arthropods).

And above all, come to Harvard, which has the largest collection of ants in the world, for your Ph.D. The following year, after Brown had departed for Australia to conduct fieldwork on that little-studied continent, Ed did transfer to Harvard. He remained there for the rest of his career, in time attaining a full professorship and the curatorship of insects, positions previously held by William Morton Wheeler, and he even inherited Wheeler's old desk, complete with pipe and tobacco pouch in the lower right-hand drawer. In 1957 he visited William Mann at the National Zoo in Washington. The elderly gentleman, in his last year as director, gave Ed his library on ants. Then he took Ed and his wife, Renee, on a tour of the zoo—past the elephants, leopards, crocodiles, cobras, and other wonders, along the fringes of Rock Creek Park, and thus, for an enchanted hour, back into the dreams of Ed's childhood. He could not have known the thrill that the closure of the life cycle gave to the aspiring young professor.

The years at Harvard were crowded with work in the field and the laboratory. The result was more than two hundred scientific publications. Wilson's interests expanded occasionally into other domains of science and even human behavior and the philosophy of science, but the ants remained his talisman and enduring source of intellectual confidence. Twenty of his most productive years on these insects were spent in close contact with Hölldobler. Sometimes the two entomologists worked separately on their own projects, on other occasions as a two-person team, but always they enjoyed nearly daily consultations. In 1985, Hölldobler began to receive irresistibly attractive offers from universities in Germany and Switzerland. When it was evident that he might actually go, he and Wilson decided to write as thorough a treatise as possible on ants, to serve as a vade mecum and definitive reference work for others. The result was *The Ants*, published in 1990, dedicated to "the next generation of myrmecologists" and replacing at last Wheeler's magnum opus of 80 years' standing. It was the surprise winner of the 1991 Pulitzer Prize in General Nonfiction, the first unabashed scientific work to be so honored.

At this time our careers had come to a fork in the road. The examination of social insects, like most of the rest of biology, had reached a high level of sophistication that required ever more elaborate and expensive equipment. Where previously rapid advances in behavioral experimentation could be made by a single investigator with little more than forceps, microscope, and a steady hand, now there was and remains a growing need for groups of investigators working at the level of the cell and molecule. Such concentrated effort is especially needed to analyze the ant brain. All of ant behavior is mediated by a half million or so nerve cells packed into an organ no larger than a letter on this page. Only advanced methods of microscopy and electrical recording can penetrate this miniature universe. High technology and cooperative efforts among scientists with different specialties are also needed to analyze the almost invisible vibrational and touch signals used by ants in social communication. They are absolutely necessary to detect and identify the glandular secretions used as signals; some of the key compounds are present in amounts of less than a billionth of a gram in each worker ant.

The University of Würzburg offered the facilities to attain this level of expertise. Martin Lindauer, his mentor, had moved there in 1973 and was now retiring. The university decided to expand the study of social-insect behavior, and asked Hölldobler to accept a chair to lead a new group in behavioral physiology and sociobiology. He chose to go, and thus, a century after William Morton Wheeler's visiting scholarship, the link between Harvard and Würzburg was reestablished. The Leibniz Prize, a million-dollar research award from the Federal Republic given to build scientific fields in Germany, was awarded to Hölldobler shortly after his arrival. The Würzburg group is now proceeding strongly into experimental studies of the genetics, physiology, and ecology of the social insects.

A different urgency propelled Wilson onto a divergent path. The muse he celebrated had always been biological diversity—its origins, quantity, and impact on the environment. By the 1980s biologists had become fully aware that human activities are destroying biodiversity at an accelerating rate. They had made the first crude estimates of this erosion, projecting that, largely through destruction of natural habitats, fully one-quarter of the species on earth could disappear within the next 30 or 40 years. It was becoming clear that in order to meet the emergency, biologists must map the diversity around the world far more precisely than ever before, pinpointing the habitats that both contain the largest number of distinctive species and are the most threatened. The information is needed to assist the salvage and scientific study of endangered forms. The task is urgent and has only begun. As few as 10 percent of the species of plants, animals, and microorganisms have received so much as a scientific name, and the distributions and biology of even this group are poorly understood. Most diversity studies depend on the best-known—"focal"—groups, in particular mammals, birds and other vertebrates, butterflies, and flowering plants. Ants are an additional candidate for elite status, being especially suitable because of their great abundance and conspicuous activity throughout the warm season.

Now as in previous years, Harvard University has the largest and most nearly complete ant collection in the world. Wilson felt an obligation beyond a natural attraction to the subject to harness the collection in the effort to make ants a focal group of biodiversity studies. In collaborating with Bill Brown, now at Cornell University, he set out to scale the Mt. Everest of ant classification: a monograph on *Pheidole,* by far the largest genus of ants, with a thousand or more species to analyze and classify. Their effort when completed will include descriptions of 350 new species from the Western Hemisphere alone.

Hölldobler and Wilson still manage to meet and collaborate in field studies once a year, in Costa Rica or Florida. There they hunt for new and poorly known kinds of ants, Wilson to add to the full measure of diversity, Hölldobler to select the most interesting species for close study at Würzburg. Meanwhile, myrmecology is rising in popularity among scientists. The eccentric tinge is gone, although the netherworld has lost none of its alien mystery.

The Life and Death of the Colony

ANT QUEENS, hidden in the fastness of well-built nests and protected by zealous daughters, enjoy exceptionally long lives. Barring accidents, those of most species last 5 years or longer. A few exceed in natural longevity anything known in the millions of species of other insects, including even the legendary 17-year cicadas. One mother queen of an Australian carpenter ant kept in a laboratory nest flourished for 23 years, producing thousands of offspring before she faltered in her reproduction and died, apparently of old age. Several queens of *Lasius flavus,* the little yellow mound-building ant of European meadows, have lived 18 to 22 years in captivity. The world record for ants, and hence for insects generally, is held by a queen of *Lasius niger,* the European black sidewalk ant, which also lives in forests. Lovingly attended in a laboratory nest by a Swiss entomologist, she lasted 29 years.

The fecundity of successful queens during these prolonged lifetimes varies greatly according to species but is always impressive by human standards. Queens of some of the slow-growing specialized predators produce a few hundred daughter workers, plus perhaps a dozen or so queens and males. Near the upper extreme, queens of the leafcutter ants of South and Central America each give birth to as many as 150 million workers, of which 2 to 3 million are alive at any given time. African driver-ant queens, possibly the world champions, may produce twice that number, a mass of daughters exceeding the human population of the United States.

Uneasy, however, lies the head that wears the crown. For every queen that starts a colony, hundreds or thousands die in the attempt. During the reproductive season successful colonies spew forth swarms of virgin queens and males that fly or crawl away in search of mates from other colonies. Most are quickly seized by predators, fall into water, or simply lose their way, later to die. If a young queen lives long enough to be inseminated, she breaks off her dry, membranous wings and searches for a place to build her nest. But the odds still weigh heavily against her. She is unlikely to find the right spot and complete the excavation before being found by predators.

The cruel lottery of colony foundation quickly becomes apparent

when you consider a representative example. Suppose that colonies last for five years and that on average only one virgin queen from five colonies will successfully start a colony each year. If a typical colony releases 100 virgin queens a year, then only one in 500 has a chance.

The males stand no chance at all. Every one dies within hours or days after leaving the mother nest. A very small number can expect to win the lottery in a Darwinian sense, even though they die in the process, by inseminating one of the rare successful queens. But the vast majority lose both their bodies and their genes. Each winning male will leave hundreds or thousands of offspring, most born months or years after he has died. The feat is accomplished by a kind of sperm bank, evolved by ants millions of years before humankind dreamed of the same technique. After receiving the ejaculate from the male, the queen stores it in an oval bag located near the tip of her abdomen. In this organ, called the spermatheca, the individual sperm are physiologically inactivated, and they can remain in suspended animation for years. When at last the queen lets them back out into her reproductive tract, either one at a time or in small groups, they become agile again and ready to fertilize the eggs passing down the tract from the ovaries.

The slaughter of failed reproductive hopefuls can be seen all over the eastern United States at the end of each summer, when the "Labor Day ant," *Lasius neoniger,* attempts colony reproduction. The species is one of the dominant insects of city sidewalks and lawns, open fields, golf courses, and country roadsides. The dumpy little brown workers build inconspicuous crater mounds, piles of excavated soil that encircle the entrance holes, causing the nests to look a bit like miniature volcanic calderas. Emerging from the nests, the workers forage over the ground, in among the grass tussocks, and up onto low grasses and shrubs in search of dead insects and nectar. For a few hours each year, however, this routine is abandoned and activity around the anthills changes drastically. In the last few days of August or the first two weeks of September—around Labor Day—at five o'clock on a sunny afternoon, if rain has recently fallen and if the air is still and warm and humid, vast swarms of virgin queens and males emerge from the *Lasius neoniger* nests and fly upward.

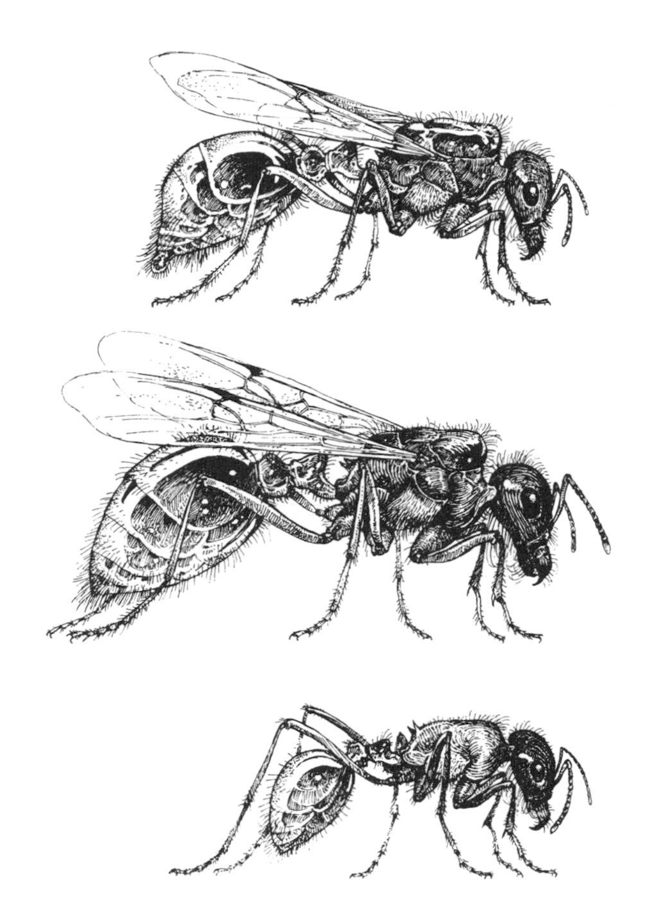

From top to bottom, a winged male, a winged virgin queen, and a worker of the American harvester ant *Pogonomyrmex barbatus.* (Drawing by John D. Dawson, courtesy of the National Geographic Society.)

For an hour or two the air is filled with the winged ants, meeting and copulating while still aloft. Many end up splattered on windshields. Birds, dragonflies, robber flies, and other airborne predators also scythe through the airborne ranks. Some individuals stray far out over lakes, doomed to alight on water and drown. As twilight approaches the orgy ends, and the last of the survivors flutter to the ground. The queens scrape off their wings and search for a place to dig their earthen nest. Few will get far on this final journey. They must pass through a terrible gauntlet of birds, toads, assassin bugs, ground beetles, centipedes, jumping spiders, and other hunters of such vulnerable prey. Most deadly of all are worker ants, including those of the ubiquitous *Lasius neoniger,* always on the alert for territorial intruders.

For ants, the nuptial flights are the supreme moment of the life cycle. Colonies may grow hungry, enemies may carry away part of the worker force, and a hundred other misfortunes may set the colony back by fractions of its capacity. Recovery still remains possible. But if the nuptial flight is missed, or if the timing is off, all the colony's effort comes to nothing. At the moment of the nuptial flight the colony becomes manic. Virgin queens and males, assisted by frantic swarms of workers, rush forth and take off. The tactics the sexes then use to accomplish mating vary from one species to the next but are always both hasty and precarious. Late on a July day in 1975, while walking across desert flats in northern Arizona, Hölldobler discovered one of the most spectacular of all examples, involving large red harvesting ants of the species *Pogonomyrmex rugosus.* On a ground the size of a tennis court, in the midst of open terrain unmarked by any distinctive physical features, were large masses of queens and males roiling on the ground. From five o'clock to dusk two hours later, winged queens flew in, were mated, and flew out again. As each one landed, she was rushed by 3 to 10 males, who struggled to mount and inseminate her. When several copulations had been achieved, the queen terminated the activity with a squeaking signal made by rubbing her narrow waist against the rear segment of the body. On hearing this "female liberation signal," the males ceased their attentions and ran away in search of another receptive female. Although most of the queens flew out of the area soon after mating, the males remained behind to continue their sexual attempts. Within a few days they died there.

Year after year Hölldobler returned in July to the same spot and always found the harvester-ant swarms active. Although the individual queens and males were forever new, having been born that very year, they nevertheless somehow found their way to the same patch of ground. This area was similar to the leks of birds and antelopes, places where males return each year to sing and display to one another and whatever females they can entice to join them. But while some of these vertebrate lekking

After the nuptial flight the newly inseminated harvester-ant queen breaks off her wings by pushing them forward with her middle legs and hindlegs. (Drawing by John D. Dawson, courtesy of the National Geographic Society.)

males are old enough to remember where to go from their previous experience, the ants are not. They must rely on instinct and special cues that emanate from the lekking grounds, signals that trigger their ancestral genetic memories. No one has yet learned how the rendezvous is accomplished, for the leks possess no obvious views, odors, or sounds that distinguish them from the surrounding terrain.

The societies of most ant species, including the American Lasius and harvesting ants, reproduce like plants. They throw out large numbers of colonizing queens, like so many seeds, on the chance that at least one or two will take root. A few, however, follow more cautious investment strategies. Queens of some of the European wood ants venture only to the surface of the home nest, linger there just long enough to be inseminated, then scurry back into the subterranean chambers. The colony later multiplies when one or more of the fertile queens, surrounded by part of the worker force, walk to new nest sites. Virgin queens of army

As the first step in colony founding, the young queen excavates a nest in the soil. (Drawing by John D. Dawson, courtesy of the National Geographic Society.)

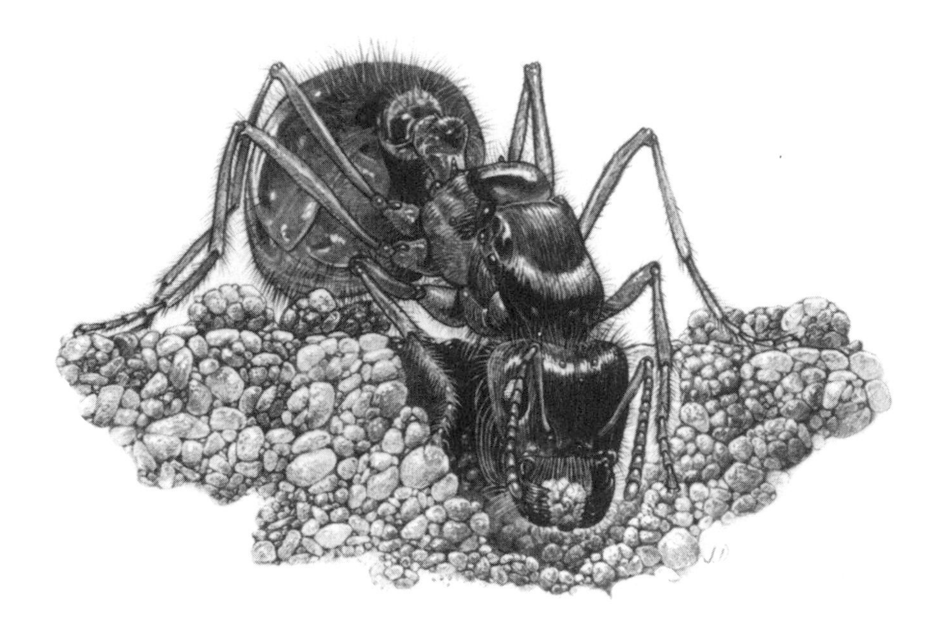

ants are even more tightly protected. Lacking wings altogether, they are simple egg-laying machines; never leaving the company of the workers, they wait for the arrival of winged males from other colonies. In one of the rare instances in which ants accept representatives of alien colonies, the workers allow the suitors into the colony long enough to mate.

The quality of ant social existence is deeply affected not just by the life cycle of the colony but by that of each colony member. The individual ant, like all other members of the insect order Hymenoptera and a majority of all insects, undergoes complete metamorphosis during its growth and development, passing through a sequence of four radically different stages: the queen lays an egg, from which hatches a larva, which grows and turns into pupa, from which the adult finally emerges. The significance of this multiple rebirth is the extreme divergence it allows between the larva and adult. The larva (caterpillar, grub, or maggot) is a feeding machine, wingless and small-brained. Its anatomy and repertory of biological responses have been designed during evolution to enlarge the body rapidly while defending the organism against enemies. The adult is an entirely different creature. Typically winged, or equipped with strong running legs, or both, it is built for reproduction and dispersal to new hunting grounds. Its diet is often different from that of the larva, centered on carbohydrates for sustained energy rather than on protein for growth. In extreme cases it eats nothing at all, living on the energy stores accumulated during the larval period. The pupa, to complete our account of the cycle, is simply the quiescent stage during which tissues are reorganized from the larval form into the adult form.

The grublike larvae of ants can do little or no work and must be nurtured much like human infants. Their reliance on the adults is increased by their limited mobility; even if they could feed themselves—and larvae of some primitive ant species have this capacity—their fat, legless bodies prevent them from traveling to distant food sources. For the same reason a large part of the effort of the adult workers must be devoted to the care of the larvae. They search away from the nest for food to supply their helpless siblings, and they protect and clean them with lavish attention. Just as the helplessness of human children binds families together and leads to many other social conventions, the dependence of the young on their adult sisters forms the core of social life in ants.

After reaching the adult stage the young queen undergoes yet another radical transformation. She changes from a highly versatile, self-reliant adult into a helpless colonial mendicant. While a young virgin still resident in her birth nest, she is ready with little notice to fly away on her own and to mate with the winged males. She alights and sheds her wings, builds a nest single-handedly, and raises the first brood of workers unaided over a period of weeks or months. Then abruptly, within a few days, the roles are switched and the workers begin to take care of her, reducing her to little more than an egg-laying machine, a demanding beggar who trails behind the workers as they move from one room or gallery or nest to another. Thus diminished psychologically, she cannot be a ruler in any overt physical sense. She issues no commands, but she does remain the prime focus of attention of the workers, whose lives are consecrated to her welfare and reproductive activity. The driving force of the relationship is Darwinian: only through the abundance of the new virgin queens, their sisters, and the replication of genes identical to their own, can the workers truly succeed.

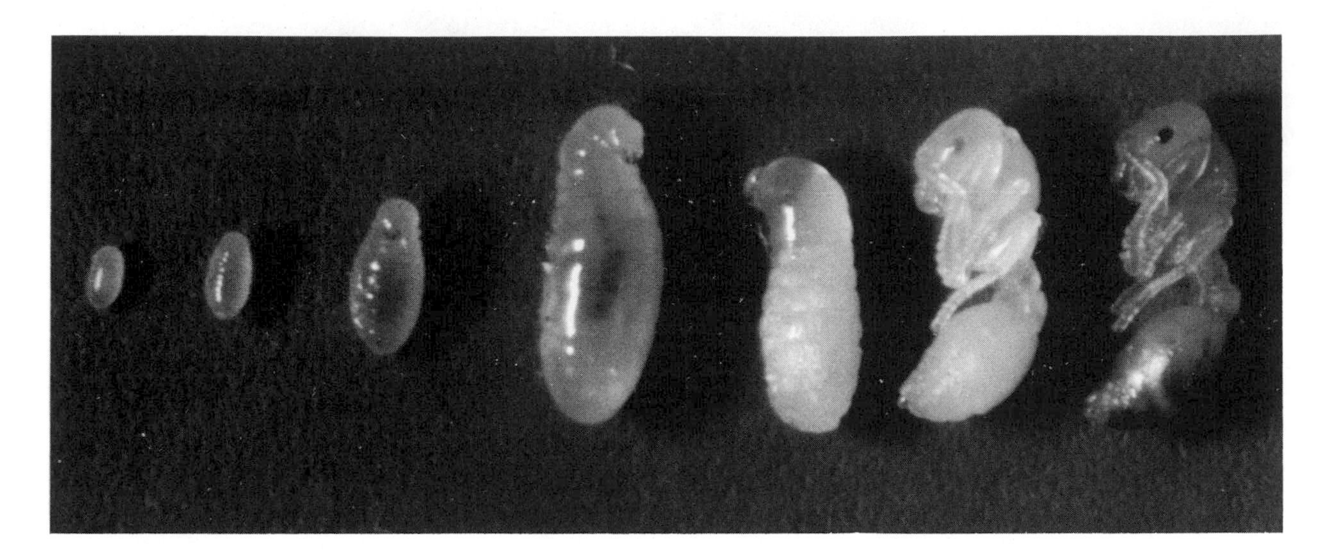

The complete development of a worker of the European ant *Leptothorax acervorum.* From left to right: egg, newly hatched (first instar) larva, half-grown larva, fully grown larva, prepupa (adult tissues forming), unpigmented pupa, and pigmented pupa ready to emerge as an active six-legged adult. (Photograph by Norbert Lipski.)

The workers of a typical ant colony are all daughters of the queen. The males, her sons, are generated after the worker population is well established and before the mating season. They live for only a few weeks or months, and with certain rare exceptions, and then only under special circumstances, they do no work. Males are thus drones in the original, Old English sense of the word: *drons,* parasites who live on the labor of others. They are also drones in the modern technological sense, flying sperm-bearing missiles constructed only for the instant of contact and ejaculation. While in the nest, however, they are totally dependent on their amazon sisters and apparently are tolerated solely for their ability to transmit the colony's genes.

The sex of ants is determined, like that of other hymenopterans such as bees and wasps, by the simplest device imaginable: when an egg is fertilized, it produces a female, and when it is unfertilized, it produces a male. This procedure allows the queen to control the sex of her offspring. By closing off the valves at the entrance of her own sperm delivery tubes, she makes sons. For most of the year, however, she holds the valve open to allow fertilization to proceed, and thus she makes daughters. In the early stages of colony development, all the daughters are retarded in their growth. They are small in size and sprout no wings. Their ovaries, if any exist at all, are relatively unproductive. And so they mature into workers, servants to the colony. Later, when the colony has grown large, some of the female larvae develop fully into virgin queens, with wings and fully formed ovaries, ready to create new colonies.

The virgin reproductives, both queens and males, are destined to depart on nuptial flights to initiate the next colony life cycle. The mother colony loses the energy and tissue put into them, but from an evolutionary perspective, they represent a crucial investment. In the language of economics, the colony decapitalizes itself in order to replicate and disseminate its genes.

What, then, guides the colony's investment? What causes a female to grow into a fertile queen instead of a sterile worker? The deciding factors are environmental rather than genetic. All the females of a colony possess the same genes with respect to caste—and any female after conception can turn into either a queen or a worker. The genes merely provide the *potential* to develop into either a worker or a queen. The controlling environmental factors are several in kind, varying among species. One is the amount and quality of food received by the larva. Another is the temperature of the nest at the time the larva is growing up. Still another is the physical condition of the queen. If the mother ant is healthy, she produces secretions during most of the year that inhibit the larvae from developing into queens. In this one category the mother deserves the name we have given her—queen, or ruler of the colony. She not only determines whether an offspring will be male or female, but also assigns caste to her daughters. Yet even here, the workers exercise a kind of ultimate, parliamentary control. They alone decide which of their growing brothers and sisters will live or die, and hence they determine the final size and composition of the colony.

The peculiarities of the life cycle and the caste system in ants arise from the fact that the colony is a family. In most species it is so tightly organized as to justify the expression "superorganism." If you look at a colony from a yard or two away, and let the image go slightly out of focus, the bodies of the individual ants seem to meld into one oversized, diffuse organism. The queen is the heart of this entity in both a hereditary and a physiological sense. She is responsible for the reproduction of the group, both the generation of the parts and the creation of new superorganisms. The ordinary lineage of colonies thus proceeds from queen to daughter queen to granddaughter queen and so on, potentially forever. The workers, the sterile sisters of each generation of virgin queens, function as little more than appendages. They are the mouth, the gut, the eyes, the whole somatic corpus of the superorganism gathered around the ovaries embodied by the queen. And while it is true that the workers make the vast majority of moment-by-moment decisions, their actions have the single ultimate purpose of allowing their mother to manufacture new queens. By this procedure, they propagate their own genes through their sister-queens.

The queen ant, then, can be viewed as an insect supported by a host of fanatical assistants, locked in deadly competition with female wasps and other solitary insects who lack the social edge. All other things being equal, this social entity, the queen ant accompanied by her workers, can be expected to prevail over solitary adversaries. Her genes will survive and spread through the world, while those of solitary competitors will decline to a corresponding degree.

If the colony exists for the welfare of the mother queen, what happens when the queen dies? It would seem logical for the workers to raise another queen to replace her. Workers are theoretically capable of creating a replacement, because some of the female eggs and young larvae still surviving can develop into queens if given the right diet. That would surely be the prudent action from the worker's point of view, because it is better to have a sister continue as queen and thus to raise nieces and nephews than to raise nothing at all. But this is not the usual course of action taken by workers when bereft of their mother; they do not follow the biologist's simple logic. In most cases the colony fails to produce a royal successor, and it declines until the last forlorn worker dies. Workers of many species possess ovaries, and while the colony is dying a few of them lay unfertilized eggs that develop into males. A sure sign that a colony is in its final days is the presence of large numbers of adult males combined with the absence of winged queens and young workers. But even this last spurt of reproduction may not occur. The workers of some species, such as the fire ants, do not possess ovaries, so that reproductive activity of the colony ceases abruptly following the death of the mother queen.

As in all of life there are instructive exceptions. Pharoah's ant, a tiny tropical species that infests the walls of human dwellings around the world, possesses queens with the shortest life span on record, only about three months. The large, diffuse colonies produce new queens steadily, which mate with their brothers and male cousins within the nest and then remain in place to join the reproductive force. With this strategy the colonies are potentially immortal. They are also able to reproduce themselves by a simple division of numbers: one group separates and walks away from another accompanied by one or more of the fertile queens. By this means Pharaoh's ants have been able to stow away in luggage and cargo and travel to far-off places—to hospitals in London and suburban homes in Chicago—and to thrive there without ever having to scatter their queens and males in nuptial flights.

Why haven't all kinds of ants bought the same ticket to colony immortality? Perhaps because the price paid is inbreeding, which entails a greatly increased risk of death and sterility. Inbred forms are also less likely to adapt to changes in the environment. Only a few species such as Pharaoh's ants live in a niche where the ecological advantage gained is more than the genetic costs paid. If this explanation is correct, we can go on to conclude that for most kinds of ants old colonies die so that new colonies may be more safely born.

Colonies with multiple fertile queens not only possess the potential for immortality but can grow to enormous size. Those of Pharaoh's ants, spreading through the walls of hospitals and office buildings, can expand to millions of workers. They compose what can appropriately be called a supercolony, an entity of theoretically unlimited size. In the north temperate zone large red and black ants of the genus *Formica* form supercolonies which live in mounds scattered across the landscape. While virgin queens usually return to one or the other of these nests shortly after mating, new nests are established by the emigration of a few fertile queens accompanied by forces of workers. The result is a vast nexus of social units that can breed and grow on their own but remain connected by the free exchange of workers along odor trails that connect the nests. One such supercolony of *Formica lugubris,* mapped in 1980 by Daniel Cherix in the Swiss Jura and still alive as we write, covers more than 25 hectares (62 acres). Its population comprises uncounted millions of workers and queens. In 1979 a supercolony of *Formica yessensis,* surely the largest animal society of any kind recorded to date, was reported by Seigo Higashi and Katsusuke Yamauchi to extend across 270 hectares (675 acres) of the Ishikari Coast of Hokkaido. It contains an estimated 306 million workers and 1 million queens, which live in 45,000 interconnected nests. As intimidating as such examples may seem, they are nonetheless quite rare in nature. Does this tell us something about the course of empire?

How Ants Communicate

WE BEGAN one of our greatest adventures, a study of African weaver ants, the day a colony of these insects took over Wilson's office. The ants were brought to us from Kenya in 1975 by two of our co-workers, Kathleen Horton and Robert Silberglied. For reasons we will explain shortly, the capture of an entire colony with a mother queen is a notable feat. Horton and Silberglied had come upon a young colony living in the branches of a small, isolated grapefruit tree, clipped off the entire nest, and bagged it without being too badly bitten. They then securely sealed the colony in a taped container and carried it to the United States with their hand luggage.

Wilson opened the box to give the ants air and placed it on a table next to the far wall of the room. Then he sat at his desk to work through correspondence and telephone calls. Two hours later, glancing past the pile of papers, he saw a scattered group of weaver ants advancing from the far side of the desk. About the size of a pencil eraser, large-eyed, and clear yellow in color, the ants were walking cautiously toward him while watching his every move.

When Wilson leaned forward for a closer look, the ants did not retreat, but challenged him back: raised their antennae to sweep the air while lifting their abdomens high and opening their mandibles wide in the characteristic threat gesture of the species. A photograph of the same posture, taken later in the wild by Hölldobler, was used for the jacket illustration of our 1990 encyclopedic book *The Ants.*

Entomologists are unaccustomed to such self-confidence, not to say arrogance, on the part of animals one-millionth their size. But sang-froid is part of the charm of the African weaver ants, known more precisely by their formal scientific name *Oecophylla longinoda.* They are bold in manner and decisive in action, and those qualities, plus their large size (for an ant) and the fact that much of their social behavior is displayed in the sunlight, where it can be easily seen and photographed, made them irresistible to us as research scientists. We seized the opportunity and undertook a careful study of the species that extended intermittently through the late 1970s and early 1980s. Our odyssey began in Wilson's laboratory and ended with studies by

Hölldobler in the field in Kenya. In time Hölldobler extended it to include the other living weaver ant, *Oecophylla smaragdina* of Asia and Australia.

In the weaver ants we uncovered some of the most complex social behavior known in the animal kingdom. Their pheromone communication system, based on chemical secretions passed back and forth to be tasted and smelled, proved to be the most sophisticated ever discovered in animals. All the many hours we spent with them were richly rewarded.

In the forests of Sub-Saharan Africa the weaver ants are among the rulers of the treetops. Their mature colonies are immense in size, containing a single mother queen and more than half a million daughter workers. In the Shimba Hills of Kenya, where Hölldobler conducted his field studies, single colonies hold territories that extend through the canopies and over the trunk surfaces of as many as 17 large trees. If human beings were similarly organized and the scale adjusted upward to account for larger body size, the weaver-ant hegemony would be the equivalent of control by a mother and her children of at least 100 square kilometers of terrain. We say at least, because the true territory of the ants is not the flat area of the forest they occupy and we conventionally measure in two dimensions, but all the vast surface of the vegetation, comprising every square millimeter of leaf, branch, and trunk from the tops of the trees down to the ground.

The weaver ants protect their domain as if it were a garrison state. They attack mammals and other intruders viciously. They hunt down members of neighboring weaver-ant colonies that intrude into their territory, and kill them. They destroy workers of most other ant species as well, together with nearly every additional kind of insect they can reach. Nearly all of their small victims are then taken back to the nest and eaten. Battles between neighboring weaver-ant colonies are so severe that they create narrow, unoccupied border corridors between the territories, a kind of "no ant's land."

The closely related non-African species of weaver ants *Oecophylla smaragdina* maintains barracks nests near the borders, in which aging workers stand guard. These individuals, no longer as capable as younger workers in the care of young, nest repair, and other domestic tasks, position themselves to be the first to meet enemies that breach the colony's territorial boundary. Near the end of their useful lives, they assume the greatest risks on behalf of the colony. It can be said that while human societies send their young men to war, weaver-ant societies send their old ladies.

To study the behavior of the African species more closely under controlled conditions, we allowed small colonies from Kenya to colonize potted lemon trees in our laboratory. Early on we noticed a peculiar habit that had escaped the notice of earlier researchers. Most ants defecate either in a remote corner of the nest or in a special garbage area outside the nest, a pile of detritus entomologists call the kitchen midden. The weaver ants are not so meticulous. They drop excrement everywhere they go. In fact they seem determined to spread their fecal odors over the full extent of their territory. When we allowed our captive colony to enter a potted tree or the paper-covered floor of an area they had not previously visited, the rate of defecation soared. At frequent intervals, far beyond what could have been their physiological need, the workers touched the tips of their abdomens (the extreme posterior ends of their bodies) to the surface and extruded large drops of brown fluid through their anuses. The material quickly soaked into the surface or hardened into shiny, shellac-like buttons. Watching these Jackson Pollock dribblings, we wondered if it was possible that the weaver ants use excrement to signal ownership of land in the same way that dogs and cats spray urine to demarcate their territories.

We tested this idea by staging experimental wars in the laboratory. We placed two weaver-ant colonies near each other, their nests connected by an arena (an open space enclosed by walls) in the middle into which either one or both could venture. Access to the arena was provided by bridges to each of the colonies that could be put in place or removed, like drawbridges to a castle. At the beginning of the experiment, we allowed workers from one colony to walk onto the arena floor and mark it thoroughly with fecal spots. After several days we took their bridge away, and removed the ants one by one and put them back into their nest. Then we allowed workers from the second colony to enter and explore the common ground. When this new group encountered the fecal spots, they hesitated and displayed in the typical weaver-ant hostile posture—mandibles open and abdomen lifted. Some ran home and recruited a small army of their nestmates. By signaling with odor trails and touch, they seemed to be shouting, "Follow me, fast! We've discovered enemy territory!" When, in contrast, scouts were allowed to explore an arena previously marked by members of their own colony, the rate of recruitment was far lower. Clearly the fecal substances exuded a scent peculiar to each colony.

Everybody knows about the home team advantage; when a game is played at home, the local team has a psychological edge over the visitors, and in a close match that is sometimes enough to gain victory. When we allowed both weaver-ant colonies to enter the laboratory arena at the same time, the scouts from both sides used their trails and other pheromones to summon large forces of nestmates, and violent battles ensued. Jaw-to-jaw combat was the rule, conducted either one on one or with two or more fighters ganging up on single opponents. The first ants to make contact with an enemy attempted to seize it by its appendages and spread-eagle it into immobility, while others throttled it and clipped off parts of its body. In ten such wars we precipitated (in our role as Olympic-sized agents provocateurs), the colony that eventually won, by driving the enemy fighters back onto the bridge and into their nest, was always the one that had been allowed to mark the arena with its scented feces before the opposing armies clashed.

The more familiar we became with the wars and daily lives of the weaver ants, the more sophisticated we found their communication systems to be. We discovered that weaver-ant workers not only guide one another to locations outside the nest but employ five different "messages" by which they specify the nature of the target. Each message is a compound of signals. A chemical substance is laid down as a trail and combined with a particular body movement, either a little dance or a touch of the antennae, whenever the trail-layer meets a nestmate. The chemicals are secretions from one or the other of two glands located next to the anus at the tip of the rearmost segment of the body. Both of these glands were new to science, discovered for the first time during our study. When a worker wishes to say, in effect, "Follow me, I've discovered

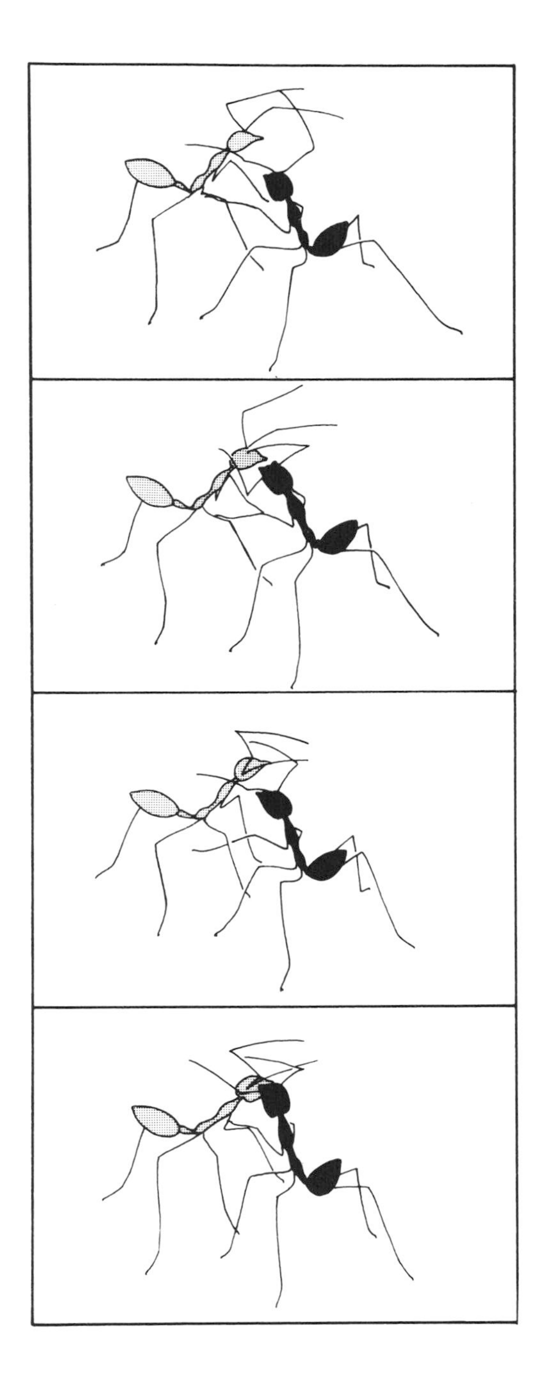

After encountering an enemy, an African weaver ant (black) recruits a nestmate by rapid back-and-forth movements. We believe this signal evolved as a ritualized form of attack behavior.

some food," she deposits a trail from one of these two secretion sources, the rectal gland, while running from the food back to the nest. When she encounters other workers, she waves her head and touches them with her two antennae. If the food is in liquid form, she opens her mandibles to offer them a regurgitated sample. The nestmate may taste the offering briefly, then run out along the trail to the newly discovered food source. A second kind of recruitment message conveys a wholly different meaning. When the scout worker locates a place where a new nest might be built, she again lays a rectal-gland trail. This time, however, she combines it with touching signals that show the other ant that she is ready to pull or bodily carry her in the direction of the new nest site. And a third message: when an enemy is encountered close by, the worker broadcasts an alarm by laying short looping trails around the intruder, drawn with substances smeared on the ground from the sternal gland, the second source of recruitment substances. In this case no special touching is employed. The two remaining recruitment signals employed by workers, using yet different combinations, direct nestmates to new, unexplored terrain or to enemies encountered at long range.

In the 1970s the British entomologist John Bradshaw and two co-workers discovered yet another alarm system in African weaver ants, this time based on multiple pheromones with different meanings. When a worker meets an enemy at the nest or on the colony territory, she releases a mixture of four chemicals from glands that partly fill the head and open to the outside at the base of the mandibles. The substances diffuse through the air at different rates and are detected by the ants at different concentrations, so that her sister workers become aware of them in stages, one after the other. Hexanal, an aldehyde, arouses the ants first, causing them to become alert. They wave their antennae back and forth in search of other smells. Then hexanol, the equivalent alcohol (hence the *o* instead of the *a*), reaches them in sufficient quantity to be detected, alarming the ants and causing them to move about in search of the source of trouble. Then comes undecanone, which attracts the workers closer to the source and stimulates them to bite any foreign object. Finally, closest to the target, they become aware of butyloctenal, which furthers the aggressive urge to attack and bite.

The weaver ants, to summarize research of the past twenty years, have come very close to employing syntax in their chemical language—the use of various combinations of chemical "words" to transmit different "phrases." They even modulate the intensity of other, primary signals composed of touch and sound.

These remarkable insects are an ancient breed. They occur in the Baltic amber of Europe as beautifully preserved fossils 30 million years old. The secret of their success in modern times, their domination of canopies of lowland forests across the Old World tropics from Africa to Queensland and the Solomon Islands, must have something to do with their efficient chemical communication. But it is based even more securely on still one more elaborate form of communication, by which the ants construct large pavilions in the tree crowns. Only with these special nests can the colonies safely shelter their huge populations. To live off the ground, most ants as large as *Oecophylla* need special cavities in the vegetation, such as the spaces beneath large pieces of peeling bark or burrows excavated and abandoned by wood-dwelling beetles. Such places are relatively few and far between, and even then rarely capacious enough to hold a colony of more than a few thousand big ants. The *Oecophylla* have surmounted this obstacle by evolving the ability to make their own housing. They weave small branches and leaves together to create large rooms with walls, floors, and roofs.

Joseph Banks was the first European to witness *Oecophylla* nest construction. While accompanying Captain Cook on his 1768 voyage to Australia, he found the weaver ant to be "living upon trees, where he builds a nest, in size between that of a mans head and his fist, by bending the leaves together and glueing them with a whiteish papery substance which held them firmly together. In doing this their management was most curious: they bend down four leaves broader than a mans hand and place them in such a direction as they choose, in doing of which a much larger force is necessary than these animals seem capable of. Many thousands indeed are employed in the joint work; I have seen them holding down such a leaf, as many as could stand by one another each drawing down with all his might while others within were employed to fasten the glue" (J. C. Beaglehole, ed., *The "Endeavour" Journal of Joseph Banks, 1768–1776,* vol. 2, p. 196, Sydney, Halstead Press, 1962).

As strange as Banks's account may have seemed to early readers, it was mostly accurate. Up to hundreds of weaver ants do line up side by side in militarily precise rows. They grip the edge of one leaf with the claws and pads of their hindlegs and the edge of the other with their jaws and forelegs, and haul the two edges together. When the gap between the leaves is wider than the length of an ant, the workers use another, even more impressive tactic, one not witnessed by Banks (who after all was busy with all the other wonders of newfound Australia): they chain their bodies together to form living bridges. The lead worker seizes a leaf edge with her mandibles and holds fast. The next worker then climbs down her body, grips her waist, and holds on. A third worker now climbs down to grip the second worker's waist, and so on ant upon ant, until chains ten workers long or more are formed, often swinging free in the wind. When an ant at the end of the chain finally reaches the edge of the distant leaf, she fastens her mandibles onto it, closing the span of the living bridge, and all the entrained force begins to haul back in an attempt to bring the two leaves together. Sometimes the gap can be closed with a single chain, but usually several such large ensembles are needed, with nestmates working side by side. Some of the workers return from the site of activity to recruit nestmates by means of odor trails. They lay the trail substances not only over the leaves and twigs but over the bodies of the ants forming the chains. Soon a living sheet of ants is formed, and it presents a startling spectacle, its surface rippling with the slight movement of thousands of legs and antennae.

But all this description leaves an important question unanswered: how do the ants decide which leaf to pull in the first place? The process, discovered by the British entomologist John Sudd in 1963, is both very simple and very effective. Ants attempting to build new pavilions, perhaps motivated by crowded conditions in the old ones, search singly along the leaf margins, stopping occasionally to pull at the edges. When they succeed at curling the leaf upward, even by a small amount, the ants hold on fast and continue pulling. This activity, marking a small measure of success, attracts other workers in the vicinity. They approach and take hold of the edge themselves. As the leaf continues to bend, still more workers converge on the site. The formula is one of simple repetition: work begets success begets added work begets still more success. A small army may be assembled before two, then three or more leaves are fastened into position by stitches composed entirely of the bodies of worker ants.

Now other weaver-ant workers move into position to apply the white "glue" described by Joseph Banks. The binding material is not a kind of paste, as he thought, but—as the German zoologist Franz Doflein discovered in 1905—threads of silk provided by the grublike larvae of the colony. How the silk is applied is the most amazing behavior of all in the repertory of the weaver ants, and the appropriate source of their vernacular name. The larvae recruited are in the final stages of development, following the last shedding of their skin as part of the growth process and prior to the next molt that will transform them into a pupa and inaugurate their changeover to the six-legged adult body form. In the nest-building process, such individuals are picked up by major workers, members of the larger of the two adult worker castes, and carried out to the leaf edges. Holding the larvae gently in their mandibles, the workers move their young charges back and forth across the leaf edges. The larvae respond by exuding threads of silk from a slit-shaped nozzle just below the mouth. Thousands of such threads stuck into place side by side spread as a whole into a sheet between the edges, in time to become a powerful adhesive that binds the leaves in place.

By their acquiescence, by allowing themselves to be turned into living shuttles, the larvae give up the silk that would otherwise be used to spin cocoons to protect their own bodies. Yet their sacrifice is not pure altruism. They are sheltered by the nest that is bound together by their bodily secretions, making it possible for them to grow more securely into adult workers, without the aid of cocoons.

To record the fine details of the spinning process we followed the entire sequence in a frame-by-frame analysis of motion pictures. The most distinctive feature of the larval behavior we witnessed, next to the release of the silk itself, is the rigidity with which the larva holds its body. There is no sign of the elaborate bending and stretching of the whole body or of the upward thrusting and side-to-side movements of the head typical of cocoon spinning in the immature stages of ants, butterflies, and other holometabolous insects. The weaver-ant larva turns itself into a largely passive instrument of the adult worker that has borne it from the interior of the nest. Occasionally the larva extends its head forward for a very short distance when it comes close to the leaf surface, apparently as a means of orienting itself just before contact. But otherwise it stays immobile and merely spins silk.

The choreography of silk spinning that ensues is a swift, precise pas de deux. The worker approaches the edge of the leaf while holding the larva in her mandibles so that the larva's head projects well out in front, as though it were an extension of her own body. The tips of her antennae are brought down to converge on the leaf edge. For two-tenths of a second the tips play along the surface, not unlike the hands of a blindfolded person feathering the edge of a table to gain a sense of position and shape. Then the worker brings the larva's head down to touch the surface. One second later she lifts it again. During this interval the worker vibrates the tips of her antennae around the larva's head, touching it lightly about ten times. The subtle tapping is apparently a signal for the larva to release the silk. We are not certain that the movement contains such a command, but while it is occurring the larva does release a minute quantity of silk, which automatically sticks to the leaf surface.

An instant before the larva is lifted from the leaf's edge, the worker raises and spreads her antennae. Then she turns her body and carries the larva directly to the edge of the opposing leaf, causing the silk to be drawn out as a thread. When she reaches this second surface, she repeats her earlier movements almost exactly. This time the larva touches the silk to the leaf and fastens the thread. Then both worker and larva return like tango dancers to the first edge to recommence the cycle. And so on metronomically, en masse, a rhythmic army of workers and larvae toils day after day, pulling together and sealing hundreds of pavilions across the great canopy empire. The ants add silken tunnels and rooms within the pavilions to create even tighter, more elaborate living quarters.

In 1964 Mary Leakey (doyenne of the Kenyan family of paleobiologists that has contributed greatly to our knowledge of the fossil history of man) sent Wilson a partial fossil colony of an extinct species of *Oecophylla* she had found during the search for early human remains. The age of the ant remains was approximately 15 million years. The fossils consisted of numerous fragments of different life stages and castes closely resembling those of the modern African and Asian weaver ants. The pupae were naked. That is, the larvae, like those of modern species, had spun no cocoons. Also, fragments of fossilized leaves were mingled with the ants. It thus appears that long ago a pavilion of weaver ants fell from a tree into a pool of water which was then covered by a rapidly congealing calcareous sediment. If that much is true, the unique social system by which *Oecophylla* weaver ants dominate the tropical canopy today appears to have been in place 10 million years before the origin of humanity.

Pheromones are favored by ant colonies in general, but the signals that bind them together are transmitted through several other channels as well. In most species, simple messages of a kind are delivered by a tapping or stroking of the body of one ant by another. The movements are simple and direct. A worker can induce a nestmate to regurgitate liquid food, for example, simply by extending her foreleg and using it to touch the other ant on a segment of its head called the labium, a part roughly equivalent to the human tongue. The response to the touch is the equivalent of the vomit reflex, except that the liquid delivered is palatable—at least to other ants—and greedily sucked up. Hölldobler found that he could trigger regurgitation simply by touching the labium of captive workers with a fine hair plucked from his own head. The ants, evidently unconcerned by his gigantic size and odd appearance, responded as though he were a friendly nestmate.

A majority of ant species also communicate by sound. They produce a high-pitched squeak by rubbing a thin, transverse scraper located on their waist against a washboard of fine, parallel ridges on the adjacent surface of the abdomen. Entomologists call this behavior stridulation. The signal is barely audible to the human ear, and then only if the ant is agitated and calling vigorously. You can just make it out by grasping a worker or a queen with a pair of forceps and bringing her close to your ear.

The squeaking serves one or another of several functions, depending on the species and on the circumstances. Some kinds of ants use it to call for help, a phenomenon first discovered in leafcutter ants of the genus *Atta* by the German zoologist Hubert Markl. During heavy rainfall, cave-ins often bury workers in parts of the labyrinthine subterranean

Two workers of an American carpenter-ant species *(Camponotus floridanus)* exchange liquid food. *Above:* the worker on the left induces regurgitation by touching her forelegs to the donor's head. *Below:* the donor, on the right, passes liquid from her crop *(K)*, the storage organ that serves as the "social stomach," through her esophagus into the mouth and crop of the recipient. Small amounts of the food are also passed from the crop into the midgut *(M)* to serve as nourishment for the donor. Waste material is passed out through the rectal bladder *(R)*. (Drawing by Turid Forsyth.)

The African weaver ant establishes vast territories in tree canopies. In the left foreground a worker threatens a member of a rival colony. Behind her, nestmates pin another enemy down, while to their right a worker runs along a branch to the nest, laying a chemical trail from the tip of her abdomen that will lead other nestmates into the fray. In the lower right still other colony members subdue a large ponerine hunting ant. (Painting by John D. Dawson, courtesy of the National Geographic Society.)

Facing page
Communication in weaver-ant colonies during territorial defense. *Above:* a worker lays an odor trail from the tip of her abdomen as a means of guiding nestmates to the enemy. *Below:* on meeting a nestmate, the recruiter performs a stereotyped "dance," raising her abdomen, opening her mandibles, and jerking her body back and forth.

Top right: alarmed weaver ants move toward the enemy, mandibles gaping and gasters cocked. *Bottom right:* on meeting an opponent, defenders spread-eagle and immobilize it until it can be killed.

Above: following the battle, weaver ants carry home both defeated enemies and their own dead to use as food. *Below:* even the swiftest and strongest invaders, such as this *Leptomyrmex* worker, can be caught and subdued.

Facing page
Weaver-ant workers form chains of their own bodies, allowing them to cross wide gaps and pull leaves together during nest construction. A simple chain is shown at the top. At the bottom, multiple chains have joined to form a powerful rope over which workers run back and forth, some laying odor trails.

Facing page

The weaver ants form rows of chains, and their combined effort creates enough force to pull stiff leaf edges together. When the leaves are in place, the ants bind them together with larval silk.

The final stage of nest construction by weaver ants is the binding of the leaves with larval silk. A worker holds a mature larva in her mandibles and moves it back and forth while the larva releases a continuous thread of silk from gland openings on the head. Thousands of such threads are aligned into a sturdy sheet.

A newly constructed nest of the African weaver ant *Oecophylla longinoda*.

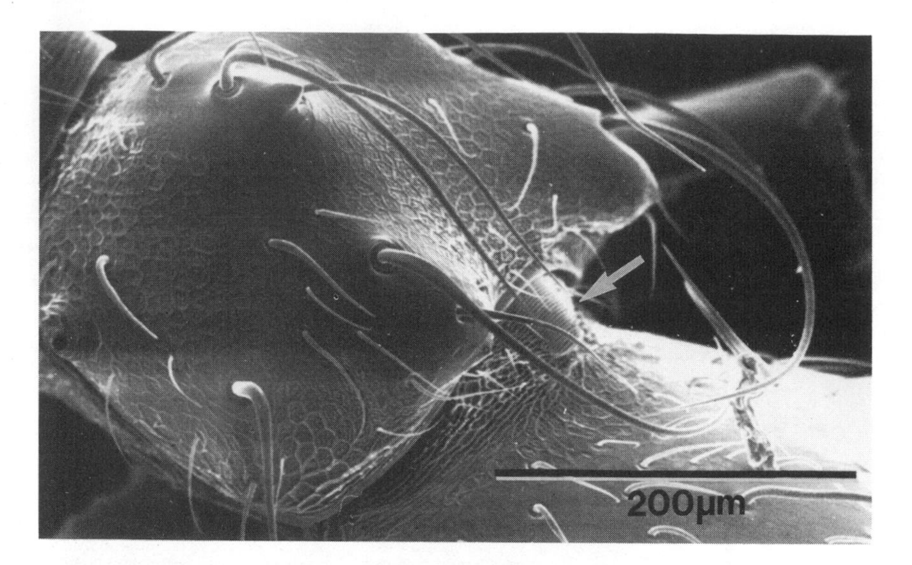

The stridulatory organ of the leafcutter ant *(Atta cephalotes)*, with which the workers make a squeaking sound to alarm their nestmates. The photograph above shows the small area, indicated by an arrow, between the postpetiole ("waist") and the gaster where the stridulation occurs. The photograph below provides a close view of the file surface of the stridulatory organ. (Scanning electron micrographs by Flavio Roces.)

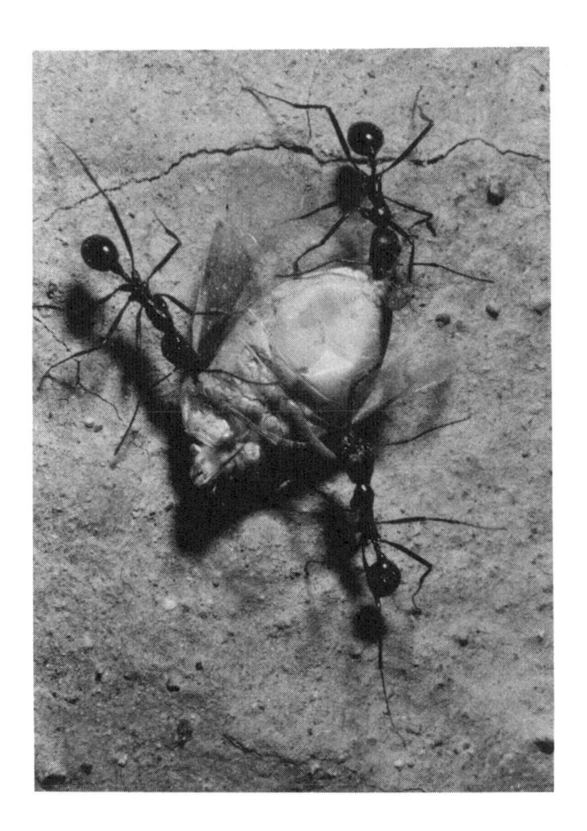

Group retrieval greatly improves the efficiency of foraging in some ant species. Here three workers of the large American desert ant *Aphaenogaster cockerelli* retrieve a dead coreid bug by swift, coordinated movements.

nests. The squeaking of the trapped ants summons nestmates to dig them out. The rescuers are not affected by that part of the sound energy carried through the air. Instead, they pick up vibrations propagated through the soil, using extremely sensitive detectors in their legs.

Recently the Argentine entomologist Flavio Roces, working with Bert Hölldobler and Jürgen Tautz in Würzburg, discovered another function of stridulation in leafcutter ants. The workers are highly selective in the vegetation they harvest. Once a forager locates a highly desirable leaf, she "sings" to others in the vicinity to join her. The vibrations produced in the stridulatory organ pass forward through the ant's body and onto the plant surface through her head, to be picked up by other workers as much as 15 centimeters away. The higher the nutritive quality of the leaf, the more intense the vibrations transmitted.

Desert-dwelling ants of the genus *Aphaenogaster* squeak for still another reason. When a foraging worker encounters a large piece of food, such as a dead cockroach or beetle, she calls to excite her nestmates. The sound is a reinforcement, not a primary signal. It does not by itself attract other workers, but instead induces them to respond more quickly to the conventional chemical recruitment signals and body touching.

Another mode of auditory communication, used for example by carpenter ants of the genus *Camponotus,* is simply to rap the head on a hard surface. The sound is transmitted through the substrate and serves to alert nestmates to danger. The species employing this device are mostly those that live in dead wood or in paperlike chambers they build themselves from chewed vegetable fibers.

Tapping, stroking, squeaking, and body-contact dancing are impressive to watch, but they are still too limited to compose a full working vocabulary. Nor can the ants rely on vision, a sense developed only sparingly by the vast majority of species, many of which stay permanently underground. In the unlighted recesses of the soil nests, in close and still air, the preferred mode of communication is by pheromones. Ants are in fact walking batteries of exocrine glands, which manufacture a large variety of such substances. We estimate that ant species generally employ between 10 and 20 such chemical "words" and "phrases," each conveying a distinct but very general meaning. Among the categories best understood by biologists who study them are attraction, recruitment, alarm, identification of other castes, recognition of the larvae and other life stages, and discrimination between nestmates and strangers. Other pheromones from the queen inhibit the laying of eggs by her daughters as well as the development of her own daughter larvae into rival queens. Others that appear to be manufactured by the soldier caste (large ants specialized to defend the colony) are also inhibitory, decreasing the percentage of larvae that grow up into soldiers. This suppression is not a selfish act by which soldiers avoid competition for their jobs. Rather, it serves the good of the society as a whole. It is a negative feedback loop that stabilizes the size of the defense force, ensuring that the other castes responsible for hour-by-hour functioning of the colony are always well enough represented to fulfill their tasks.

Workers of the subterranean ant *Acanthomyops claviger* alert their nestmates to danger by discharging mixes of chemicals from two sources, the mandibular gland *(M)* and Dufour's gland *(D)*. These substances, which are toxic, are also used to repel enemies. (From F. E. Regnier and E. O. Wilson, *Journal of Insect Physiology,* 14, no. 7: 955–970, 1968.)

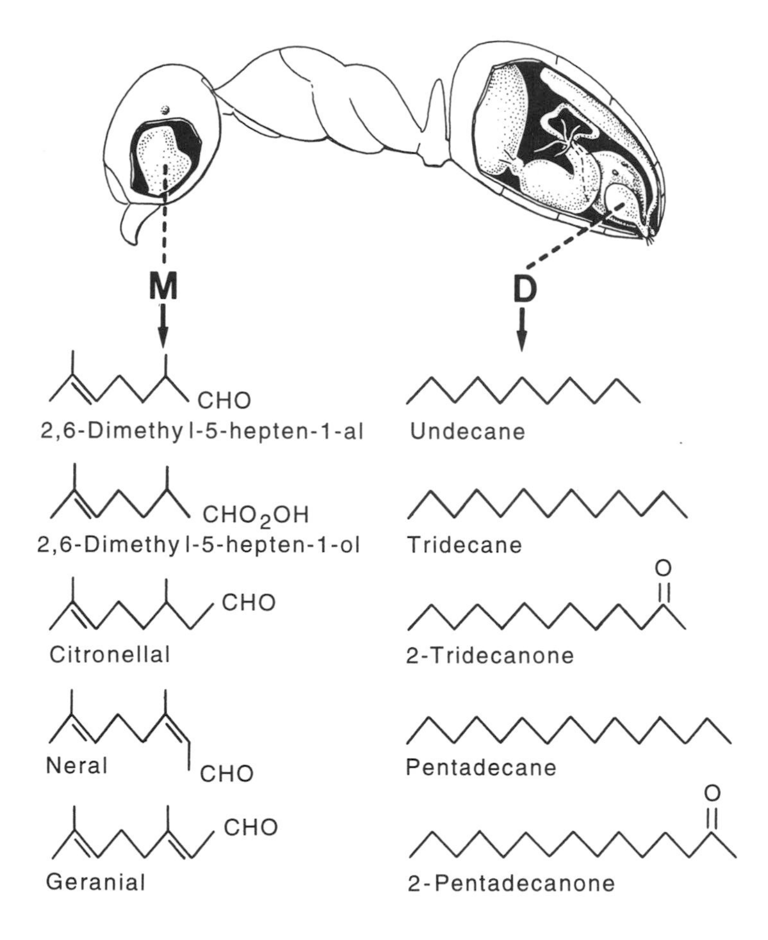

The prevalence of communication by chemicals presents no deep intellectual mystery. Its initial strangeness to humans is due simply to our own physiological limitations. We are microsmic, able to tell only a few odors apart. Our vocabulary contains only a few words to express our sensations: sweet, fetid, stringent, sour, musky, acrid . . . and a few more, and then we have exhausted the list and rely on allusions to specific objects otherwise recognized by their visual properties, such as coppery, rose-scented, banana-like, cedarish, and so on. In sharp contrast we are superb in the auditory and visual channels, upon which we have built our civilizations. Ants have followed another evolutionary

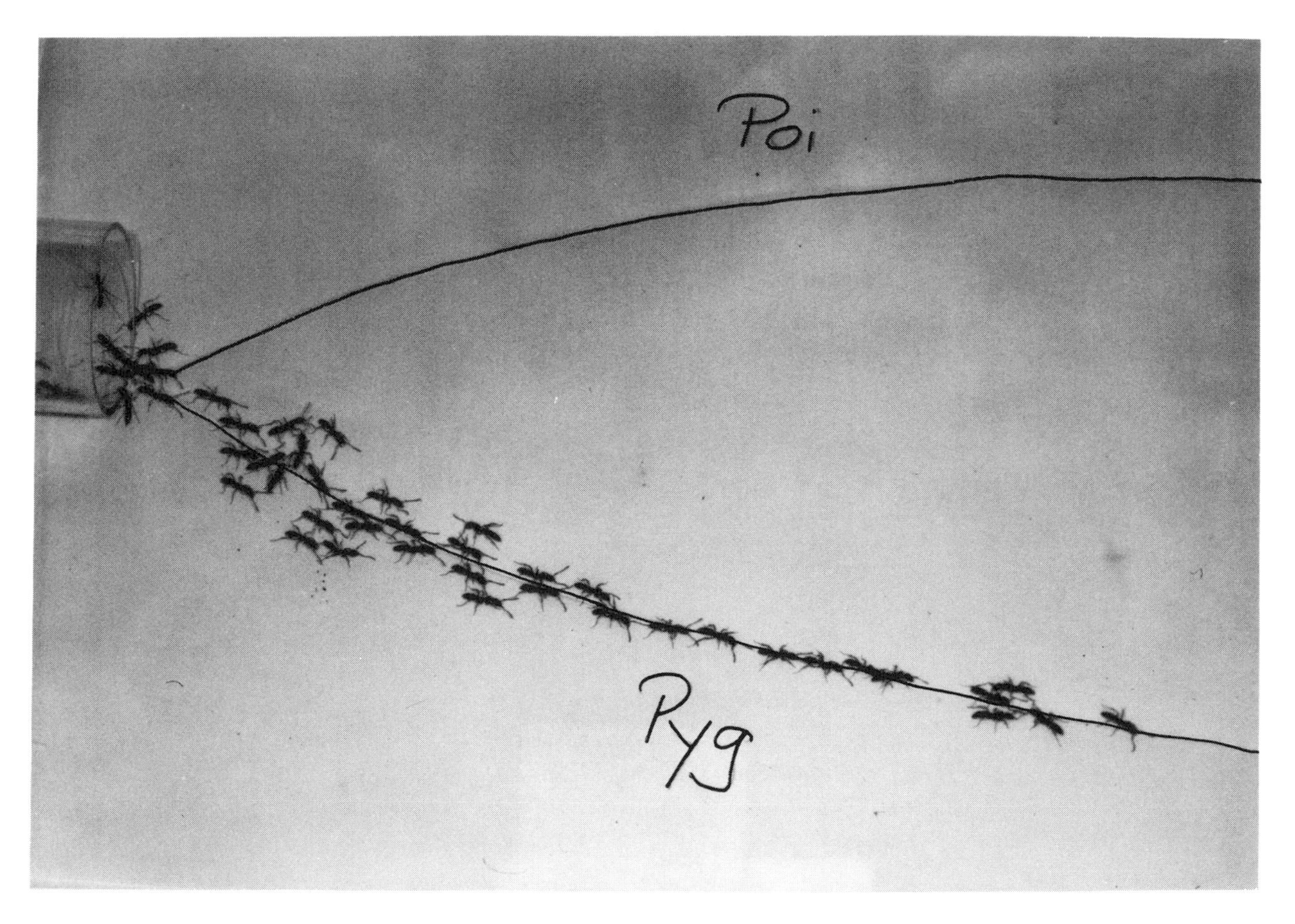

Laying artificial trails is a standard method of determining the potency of the natural secretions produced by the ants, in this case an Australian species of *Leptogenys.* The worker produces trail pheromones in both her poison and pygidial glands. The pygidial-gland substances are the more effective: when artificial trails of both pheromones are brushed along pencil lines, all the workers follow the pygidial-gland trail *(Pyg).* Other tests showed that trails made from the poison gland *(Poi)* serve mainly for orientation once the ants have been alerted by the pygidial-gland pheromone.

course. They accomplish relatively little with sound and virtually nothing with sight.

When we peer down on their colonies from our preposterous height, like huge movie monsters looming above a city, we can at first make little sense of the mode of organization. We see ants scurrying around as though blindfolded and sworn to silence. Their rules of organization were a mystery until chemists joined biologists in the effort to identify the minute traces of organic compounds by which the insects communicate. Each worker typically carries no more than millionths or billionths of a gram of each of her pheromones at a given moment, too little in most cases for the human nose to detect.

Yet we should not think the ants peculiar, at least not when compared with the remainder of life. The vast majority of species, more than 99 percent if microorganisms are included, communicate mostly or entirely with molecules. By necessity single-celled creatures evolved to respond to subtle chemical changes in their environment, including the smell of approaching predators, prey, and potential mates. Their microscopic bodies were equipped to read chemical compounds accurately—but not patterns of light and sound. As larger organisms appeared, the cells composing their tissues continued to communicate by hormones, which are molecules that travel from one part of the body to another as chemical messengers. Hormones mediate physiological reactions to keep tissues and organs well coordinated. Only insects and other animals that are very large relative to microorganisms have enough cells to create eyes and listening devices capable of processing complex information. And only with such an expanded capacity can organisms communicate efficiently in the audiovisual channel. Ants did not venture into the sensory world we occupy. Instead they remained masters of a more ancient craft. The vertebrate animals that gave rise to mammals were the ones that veered off the traditional evolutionary pathway. They entered a new sensory domain from which we now, at last, can take the measure of the larger world that both they and we occupy.

War and Foreign Policy

THE SPECTACLE of the weaver ants, their colonies locked in chronic border skirmishes like so many Italian city-states, exemplifies a condition found throughout the social insects. Ants in particular are arguably the most aggressive and warlike of all animals. They far exceed human beings in organized nastiness; our species is by comparison gentle and sweet-tempered. The foreign policy aim of ants can be summed up as follows: restless aggression, territorial conquest, and genocidal annihilation of neighboring colonies whenever possible. If ants had nuclear weapons, they would probably end the world in a week.

Citizens dwelling in cities and towns along the Atlantic seaboard from Bangor to Richmond walk by and sometimes inadvertently stamp out ant wars many times each summer, usually without taking notice. On a bare patch in a lawn, at the edge of a sidewalk, or in a gutter can often be seen, if the eye is allowed by habit to travel to the ground, masses of the pavement ant *(Tetramorium caespitum)* the size of a man's outstretched hand. Examined closely, preferably with a magnifying glass, the dark splotches are found to comprise hundreds or thousands of worker ants gripping one another in mandible-to-mandible combat, pulling, strangling, and cutting off the limbs of opponents. The belligerents are members of rival colonies engaged in a territorial war. Lines of workers run back and forth from the nests to the killing ground. The larger colony, the one able to field the largest fighting force, will as a rule either push the smaller colony into a smaller space or destroy it altogether.

Ants of many kinds conduct war with colonies of both their own species and alien species. Some employ strategies that might have been prescribed by Carl von Clausewitz, the Napoleonic Era master of battlefield science. Wilson discovered one of the most elaborate examples when he juxtaposed colonies of two species from the southern United States, the imported fire ant *Solenopsis invicta* and the abundant woodland ant *Pheidole dentata.* The fire ants are deadly enemies of the *Pheidole.* Their colonies are a hundred times larger, and if allowed to attack in confined laboratory conditions, they quickly destroy and eat the *Pheidole.* Yet *Pheidole* colonies thrive in

large numbers all around fire-ant nests in the open pine woods and thickets where both species live. How do they manage to avoid such a formidable enemy?

The secret of *Pheidole* defense is the possession of a specialized soldier caste and a three-stage strategy evidently designed to thwart attacks by fire ants. The soldiers have outsized heads encasing tiny brains hemmed in on all sides by powerful muscles that operate sharp, triangular mandibles. The soldiers do not attempt to sting or spray poisons on their opponents, the preferred offensive tactics of the great majority of ants. Rather, they work their mandibles like wire clippers, snipping off the head, legs, and other body parts of the enemy insects. The fighters, which make up only about 10 percent of the colony population, stand or walk about idly within the nest if no threat is present. Sometimes they accompany the small-headed "minor" workers on outside forays, where they help protect large food finds from takeover by other colonies. But most of the time they just sit and wait, rather like fully fueled interceptor jets on a carrier deck. Both they and the minor workers are fine-tuned by instinct to respond to fire ants more than any other enemy. The minor workers constantly patrol the ground around the nest, mostly looking for food but always alert to the approach of enemies—especially fire ants. One fire-ant worker straggling close by is enough to trigger a violent response. The *Pheidole* minor worker encountering it rushes forward in a brief attack, close enough to touch it and acquire some of the enemy odor on her own body. Then she breaks away and runs back to the nest. During the homeward run she repeatedly touches the tip of her abdomen to the ground in order to lay an odor trail with substances from the poison gland. Along the way, the scout rushes up to every individual she meets, then breaks away and continues on to the nest. Alerted by the combination of the pheromone and the trace of enemy odor carried on the scout's body, both minor workers and soldiers run out along the trail in search of the fire ant. After a brief contact with the enemy, some of the minor workers return to the nest to recruit other colony members, while the soldiers circle about and attack relentlessly. If only a single intruder is present, it is dispatched immediately. Even several fire ants can be hunted down within minutes. But total victory is not enough to satisfy the *Pheidole* soldiers. For an hour or two they search the surrounding area for additional intruders. The result of this zeal is that fire-ant scouts seldom make it home. And without dispatches from the front, their colonies are blinded. If somehow alerted to the presence of the *Pheidole* colony, the fire ants could quickly overwhelm it. The hair-trigger reaction of the defenders, however, renders the *Pheidole* undetectable most of the time.

Even when, on occasion, fire-ant scouts penetrate the protective shield and mount a full-scale campaign, the defenders have effective fallback measures. As more and more fire ants arrive at the battle scene along odor trails of their own, the *Pheidole* soldier force also continues to grow. The assembled fighters dash about in a frenzy of search-and-destroy. Fewer *Pheidole* minor workers join the action, and most of those already engaged pull away and return home. The ground is soon littered with the bodies of *Pheidole* soldiers crippled or killed by fire-ant venom, mingled with pieces of fire-ant bodies chopped off by the defenders' mandibles. In time the *Pheidole,* now greatly outnumbered, begin to fall back toward the nest. As they retreat the soldiers employ a tactic that would have been approved by Clausewitz. They close ranks to create a shorter perimeter around the nest entrance, from which they sally out into the advancing swarm of the enemy.

Meanwhile inside the nest, the minor workers prepare for a final desperate maneuver. Energized by the approaching fire-ant invasion, more and more minor workers run through the nest chambers and galleries laying odor trails and exciting their nestmates. The activity mounts rapidly. It is one of the few positive feedback actions recorded in the annals of animal behavior. The buildup culminates in an explosive response: for a few frantic minutes minor workers, many carrying eggs, larvae, and pupae in their mandibles, run out of the nest through the melee of battle and beyond to safety. There is no coordination apparent in the break-out. For once in the life of the colony, it is every ant for herself. Even the queen runs alone.

The *Pheidole* soldiers remain true to their caste. They do what they are programmed to do: stay on and fight to the death. They are the insect equivalents of the Spartan defenders who held fast and died at Ther-

Enemy specification by the ant *Pheidole dentata* (black), in which the workers respond much more aggressively to fire and thief ants of the genus *Solenopsis* (gray), than to other kinds of ants. After contacting fire-ant workers near the nest, minor workers of *Pheidole* run back and forth to the nest, dragging the tips of their abdomens over the ground to lay odor trails (depicted in the upper left). The trail pheromone attracts both minor and major workers to the battleground. The majors are especially effective in destroying the invaders, which they chop to pieces with their powerful, clipperlike mandibles. Some of the *Pheidole* are themselves crippled or killed by the venom of the fire ants. (Drawing by Sarah Landry.)

mopylae before the Persian hordes, to be commemorated there by a metal plate inscribed, "Stranger, if you see the Lacedaemonians, tell them that we lie here faithful to our instructions."

Finally, when the fire ants abandon the nest site, the *Pheidole* survivors straggle back to resume their communal life. If left undisturbed for a month or two, they are able to raise a new crop of soldiers and continue their previous life, as though nothing had happened. No revenge raids against the fire ants are undertaken. The clockwork societies of ants do not work in this humanly logical fashion.

Warfare among the ants is all about territory and food. In northern Europe, huge colonies of *Formica polyctena,* a species of wood ant, conduct cannibal wars against other colonies of their own species. The raids reach a peak in times of food shortage, especially during the start-up period of colony growth in the early spring. The Formica also attack other species of ants; on these occasions the wars are vicious enough to result in the victims' complete local annihilation. The "little fire ant" *Wasmannia auropunctata,* famous for its dense populations and painful stings, is capable under some circumstances of wiping out entire ant faunas over large areas. Accidentally introduced by commerce into one or two islands in the Galápagos Islands in the late 1960s or early 1970s, it has spread outward across the archipelago, forming in many places a living blanket of ants that kill and eat nearly all other ants in their path.

Two ant species, *Pheidole megacephala,* a native of Africa, and the "Argentine ant," *Linepithema humilis* (formerly called *Iridomyrmex humilis*), originating from southern South America, are notorious as destroyers not only of other ants but of whole native insect faunas. When *Pheidole megacephala* was accidentally brought to Hawaii in cargo ships during the last century, it multiplied prodigiously across the lowlands, devastating native insects of all kinds and probably contributing to the extinction of some native birds. It is not surprising, then, to find that these two global menaces, *Pheidole megacephala* and *Linepithema humilis,* are wholly incompatible where they meet. *Humilis* is the usual winner on subtropical to warm temperate land between 30° and 36° latitude both north and south, while *megacephala* is the winner in the intervening tropics. Because of its temperature preference, *humilis* is more familiar to dwellers of temperate zones. It dominates disturbed habitats in southern California, Mediterranean countries, southwestern Australia, and the island of Madeira. On Hawaii it occurs only above about 1,000 meters (3,300 feet) of elevation, a zone cool enough to favor it over *megacephala.* Both species penetrate new environments on foot. Like the old Zulu bands, raiding columns of workers clear the way for the pioneer communities of workers and queens, which then flow into the freshly opened nest sites and consolidate control over the surrounding terrain. New populations, by contrast, usually grow from little groups of workers and queens that stow away in cargo and baggage.

Once in a great while an ant species dominates the environment enough to challenge even human occupation. In the early 1500s a stinging ant appeared in huge enough numbers on Hispaniola and Jamaica nearly to cause abandonment of the early Spanish settlements. The colonists of Hispaniola called on their patron saint, St. Saturnin, to protect them from the ant, and they conducted religious processions through the streets in order to exorcise it. What appears to have been the same species, later given the formal scientific name *Formica omnivora,* multiplied to plague proportions on Barbados, Grenada, and Martinique in the 1760s and 1770s. The legislature of Grenada offered a reward of £20,000 to anyone who could devise a way of exterminating the ant, but without success. Left largely unchallenged, the species simply subsided on its own over a period of years. It now appears that *Formica omnivora* was the native fire ant *Solenopsis geminata,* which can be found today living peaceably as a member of most West Indian insect communities.

The tactics of battle employed by ants are extremely diverse. A few stretch the limits of insect mental and organizational capacity. In the Arizona desert a tiny, fast-moving ant, *Forelius pruinosus,* uses poisonous secretions to intimidate and steal food from honeypot ants, members of the genus *Myrmecocystus,* despite the fact that their victims are more than ten times larger. They also occasionally prevent the honeypot ants from leaving the nest altogether by gathering in hordes at the nest holes and using their chemical weapons to drive the big ants underground. The honeypot ants are thus cleared from the hunting areas around the nests, allowing the *Forelius* to harvest a larger share of the available food.

A bizarre variation of the nest-blocking technique is practiced by another malodorous little ant of the southwestern deserts, *Conomyrma bicolor.* Using chemical trails dispensed from a paddle-shaped gland at the tip of the abdomen, scouts recruit their nestmates in large numbers around the honeypot-ant nest holes. The besiegers employ chemical weapons like those of the *Forelius.* But they also pick up pebbles and other small objects with their mandibles and drop them down the vertical entrance shafts. Although no one knows exactly how the stone-dropping alters the behavior of the honeypot workers within the target nests, the effect is to reduce the amount of outside foraging they attempt. With the enemy bottled up, other *Conomyrma* workers are able to hunt for food without interference. The *Conomyrma* technique has an added interest for biologists: it is one of the rare instances of tool use among animals.

The European thief ant *Solenopsis fugax* employs a chemical mace when it invades the nests of other ant species to prey on their brood. In addition the workers are expert sappers. They first excavate an elaborate system of subterranean tunnels from their own nest to that of the target colony. The first to break through run back to their own nest and recruit a large force of nestmates. The little army pours into the enemy nest and carries away the brood, which they later consume. The invaders are able to overcome much larger ants by discharging a highly effective and long-lasting repellant substance from their poison glands. The secretion confuses and disables the opponents, allowing the thief ants to rob at will.

Another specialized form of aggression, food robbing, or kleptobiosis, occurs in a few ant species. Hölldobler studied it for many summers in the deserts of Arizona. The victims are ants of the genus *Pogonomyrmex,* which survive primarily by harvesting seeds and other edible plant material. They also gather termites intermittently, especially when these insects appear in large numbers on the soil surface following rain. The robbers are honeypot ants of the genus *Myrmecocystus,* which live on insects and sugary secretions obtained from nectaries (glands on flowers and other parts of plants) and from homopterous insects. The honeypots, which are fast and agile, often stop and inspect laden harvester ants, sometimes acting alone and sometimes in small robber packs. If the

The use of chemical weapons to repel competitors at food sites. *Above:* workers of the fire ant *Solenopsis xyloni* defend the severed abdomen of a honeypot ant by lifting their abdomens and extruding their stings to release an odorous venom. *Below:* a similar tactic is employed by foragers of *Meranoplus* defending the abdomen of a cockroach.

harvester is carrying plant material, it is allowed to pass; if its burden is a termite, it is robbed. When the harvester ant lunges at its tormentors in an attempt to bite them, the quick-footed honeypots simply run away.

The ultimate sacrifice in public service is to destroy enemies by committing suicide in defense of the colony. Many kinds of ants are prepared to assume this kamikaze role in one way or another, but none more dramatically than workers of a species of *Camponotus* of the *saundersi* group living in the rain forests of Malaysia. Discovered in the 1970s by the German entomologists Eleanore and Ulrich Maschwitz, these ants are anatomically and behaviorally programmed to be walking bombs. Two huge glands, filled with toxic secretions, run from the bases of the mandibles all the way to the posterior tip of the body. When the ants are pressed hard during combat, either by enemy ants or by an attacking predator, they contract their abdominal muscles violently, bursting open the body wall and spraying the secretions onto the foe.

About the time the Maschwitzes discovered the exploding *Camponotus*, Hölldobler stumbled upon what may be the most elaborate of all aggressive strategies in the social insects. At least one species of honeypot ants, he discovered, goes far beyond combat in territorial conquest and defense. The workers of this species, *Myrmecocystus mimicus*, rely heavily on surveillance, propaganda, and what can be termed without exaggeration a rudimentary form of foreign policy. They probe enemy territory, set up pickets, and try to bluff their foes into submission with elaborate displays that threaten but only rarely proceed to actual combat.

The martial conventions of the *Myrmecocystus* were revealed by a research technique commonly used by field biologists that is highly effective, and worth pausing here to examine in more general terms. There are two schools of field biologists recognizable, differentiated by the approaches they take to the organisms chosen for study. Members of the first group, the theoretician-experimentalists, conceive of an interesting problem that might be solved by exploring the natural environment. They subscribe to the belief that for every problem in biology, there exists an organism ideally suited to its solution. They might start, for example, by asking whether emigration plays a key role in limiting local populations. The next step is to identify a species prone to emigration—

let us say, the meadow vole. Populations of voles in the natural environment can then be fenced off, so that very little is changed except that emigration is no longer possible. Other, nearby populations can be left unfenced to serve as controls.

Researchers belonging to the second school, the naturalists, proceed in exactly the opposite direction. For every kind of animal, plant, or microorganism, they think there exists a problem which that organism is ideally suited to solve. Naturalists select particular organisms because they like to study them. Their motives are often no more complicated than that. They go into the field to learn as much as possible about the biology of the organisms, then sometimes they use the new information to search for some problem of general scientific interest. In the midst of the study of voles, for example, a particular naturalist might notice that young individuals tend to emigrate when populations become crowded. This observation leads him to guess that emigration controls population density. He might then set up a fencing experiment to test the idea.

Naturalists are opportunists. They love not merely the subject but the whole idea of the subject. Their primary aim is to learn as much as possible about all aspects of the species that give them esthetic pleasure. Organisms are their totems, to be venerated and put to the service of science. Both of us belong to this second school of biology. We are professional naturalists, and a large part of our careers has been devoted to bringing ants into the mainstream of biology.

Such was the spirit with which Hölldobler walked the desert floor near Portal, Arizona. At that time, in the 1970s, he briefly examined every species of ant he encountered, hoping to discover new phenomena. One day he saw workers of the honeypot ant species *Myrmecocystus mimicus* attacking termites while simultaneously threatening other honeypot ants. That is what he set out to study, and so our story resumes.

Honeypot workers prey on insects and other arthropods of many kinds. They are especially attracted to termites. When a scout meets a foraging party of these insects, most commonly beneath a fallen tree branch or piece of dried cow dung on which the termites are feeding, she runs back to her nest, laying an odor trail. The attractive substance in the trail is contained in rectal liquid, which is deposited through the anus in

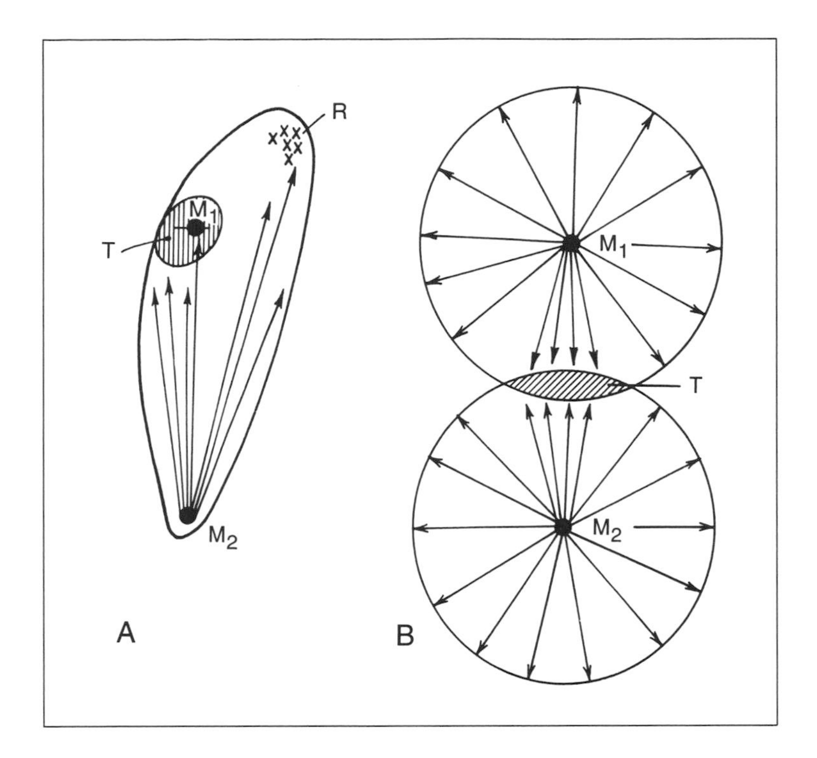

Territorial wars and expansion occur frequently in American honeypot ants *(Myrmecocystus mimicus)*. In one such episode, depicted in panel *A*, some workers from nest M_2 retrieve food from site *R*, while others engage workers from nest M_1 in a display tournament *(T)* close to the nest of these rivals, thus interfering with their ability to forage. In panel *B*, the routes of M_1 and M_2 foragers are depicted as they search in many directions around their nests for food. Overlap in foraging often leads to tournaments and even extirpation of one colony by another.

streaks and spots on the ground. The recruiter also stops to jerk her body against that of any nestmate encountered along the way. These combined signals—trail odor plus bodily contact—are enough to attract a small foraging party to the termite site. If the newly arriving recruits also find the nest of another honeypot colony nearby, some of them run swiftly back to the nest while laying recruitment trails of their own. Their action attracts an army of 200 or more workers to the vicinity of the foreign nest. Most of the recruits immediately confront the honeypot residents and attempt to keep them inside the nests, while others in the army continue to capture termites and carry them home.

The honeypot ants rarely engage in physical combat ending in injury and death. Instead they use a tournament, an elaborate display in which workers from both colonies attempt repeatedly to intimidate and chase away their rivals. The ants challenge one another back and forth across the disputed field in the manner of medieval knights, one on one. They walk about with legs stretched out in a stiltlike posture while lifting their heads and abdomens and occasionally inflating their abdomens to a slight degree. The total effect is to make each ant appear larger than it really is. Workers heighten the illusion further by climbing on top of little pebbles and clods of earth, from which they display down to their opponents. When two antagonists first meet, they perform a formicid pas de deux: they turn their bodies about to face one another head-on, then stand side by side while straining to raise their bodies ever higher, and then often circle each other slowly while drumming their antennae on the opponent's body and kicking out at her with their legs. Occasionally one leans against the opponent as though in a half-hearted attempt to push her over. All this effort is ritualized and gentle, far short of the ants' fighting potential. Either ant could easily seize and slash the opponent with her sharp mandibles, or spray her with formic acid, both actions having a fatal result. But during the tournaments such violence rarely occurs. After several seconds one of the displayers yields, and the encounter ends. The two ants then strut off on stilt legs in search of other rivals. Whenever one meets a nestmate instead of an enemy, she checks its odor with a sweep of the antennae, jerks her body up and down in a kind of salute, and walks on.

The entire bloodless performance resembles the "nothing fights" of the Maring tribe of New Guinea, in which combatants line up on either side of the territorial boundary to display their ceremonial dress, facial adornment, numerical strength, and weapons. Dances are performed and threats shouted back and forth across the field. The warriors fire arrows until someone on one side or the other is injured or killed, whereupon both parties return home. The desired result is the communication of fighting ability. All-out war is rare.

The honeypot ants, skilled communicators in a violent world, also maintain a long-term balance of power with minimum loss of blood—or (to use the correct entomological term) loss of hemolymph. Certain colonies, especially those with the greatest number of large-sized workers, the major caste, dominate neighbors, bluffing them away from the tournament arenas and into smaller foraging zones closer to their nests.

Somehow the individual combatants are able to assess the strength of the enemy and use that knowledge to respond by choosing either greater boldness or deference. How can a simple insect make such an evaluation, and arrive at such a decision? It was obvious to Hölldobler that the ants can never get a bird's-eye view of the whole arena. They are unable to count the numbers of combatants from the two colonies. Not even the entomologist can accomplish that, at least not without using stop-action film analysis conducted across the whole course of the engagement. Hölldobler proceeded to do just that, tracking scores of honeypot-ant contests in the desert. He took the data to the theoretical biologist Charles Lumsden, and together they studied the problem. Eventually they came to realize that there are at least three ways the ant workers might *indirectly* assess enemy strength. They can "count heads" while shifting from one combatant to another. If their nestmates outnumber the enemy—say three to one—they will be subjectively aware of the imbalance in their favor and more inclined to press forward. If the reverse, they will retreat. A second method is to poll the enemy. If a high percentage of the alien workers encountered are majors, the other colony is probably large, because majors are produced in high numbers only when colonies approach maturity. The third technique available to the individual ant is to judge how much queuing there is, and thus the time it takes to find an opponent who is not already engaged in ritual combat. If opponents are easily found, so that the worker is kept busy in her one-on-one displays, the enemy force is probably much larger than the army of nestmates. If the wait for an opponent is long, the enemy is weak.

After many days in the desert and at home in front of a stop-motion movie projector, Hölldobler concluded that to some degree all three measurement techniques are used by the *Myrmecocystus* fighters. He also found that very small, immature colonies, the ones most likely to be quickly defeated in a real fight, probably concentrate on the caste-polling method. That technique gives them the quickest indication of whether the opponent force is relatively large, so that they can make a rapid and sensible retreat.

In addition to the tournaments as a whole, the honeypot colonies use

Warfare in honeypot ants. *Above:* workers of a young (three-year-old) colony encounter a large worker from a mature enemy colony. *Below:* after brief aggressive displays they attack the adversary. If they are able to repel or kill her, they may forestall a raid from the enemy colony at least long enough to cover up the territorial scent marks that betray the presence of their own nest, and then close the nest entrance.

yet another method of hostile reconnaissance and limited engagement. They place guard contingents in boundary areas between nests where tournaments most often occur. The sentinels, sometimes only a few workers—rarely more than a dozen—stand for hours in the stilt-legged posture on the top of small pebbles and clumps of earth. Similar squads from neighboring colonies show up to pose in the same places, and the two forces often engage in mini-tournaments. The resulting standoff can last for days or weeks. But if the number of guards from one colony suddenly increases, the guards from the other colony run home to recruit an army of nestmates, and the confrontation escalates into a full tournament.

Do not conclude from this account, gentle reader, that honeypot ants are civilized. They possess deadly mandibular teeth and chemical weapons only thinly concealed by the formal ceremonies of the tournaments. When one colony proves decisively stronger than a neighbor, more exactly when it can summon an army about ten times more numerous than that of the opponent, the tournaments come to an end and an all-out battle breaks out. The ants bite, choke, and spread-eagle each another, until finally the larger force fights its way into the nest of the opponent, crippling and killing workers who stand in its way. They kill the queen and capture the larvae, pupae, and youngest adult workers. They also drag workers of the honeypot caste back to their own nest. These large individuals (on which the vernacular name of the whole species, honeypot ant, is based) have abdomens swollen with sugary plant secretions. They serve as living storage receptacles for the rest of the colony, regurgitating the sweet liquid to their nestmates when food is scarce. If captured, they are not killed but are incorporated into the conquering colony; there they are adopted as full members, not treated as subordinates or given menial tasks.

Yet there is no denying that the captives have lost their queen, the mother of their own colony, and that without her egg-laying capacity, they have surrendered their Darwinian raison d'être. They can no longer raise sisters, which is the main reason for belonging to a colony in the first place. The lives of ants, we are reminded by the details of honeypot-ant foreign strategy, are but little pools of harmony in an unforgiving world.

The Ur-Ants

IN 1966 the missing link of ant evolution, the Ur-ant that joins the modern forms to their ancestors among the wasps, was finally discovered. The fossil specimens composing the link provided us with proofs of some predictions we had made earlier from evolutionary theory, mixed with some exciting surprises. Prior to this find, there had been mostly frustration. The known fossil record had stopped cold in Eocene sediments some 40 to 60 million years old; earlier rocks and amber pieces seemed to offer no clues. The few specimens from the earliest, Eocene, record at the disposal of myrmecologists were poorly preserved but clearly belonged to modern groups. They were not much different in anatomy from living forms and offered no clues as to how ants came into existence.

It was known that by Oligocene times, 25 to 40 million years ago, ants had proliferated worldwide to become one of the most abundant groups of insects. Thousands of beautifully preserved specimens had been recovered in the Baltic amber of northern Europe, a fossilized tree resin of transparent, gemlike quality. As the resin ran and dripped long ago from wounds in trees, it covered swarms of the insects, representing a diversity of species, and quickly preserved many of them. By cutting and polishing the amber pieces, it is possible today to study these ancient forms in microscopic detail. The exoskeleton, the outer shell of the body and all there is to see even in living ants without resorting to dissection, was often preserved with no distortion. Fine details of the teeth, hair, and body sculpturing can still be measured through the glassy amber to the nearest hundredth of a millimeter. The specimens, it must be added, look as though they are entire bodies, but are in fact mostly rotted-out cavities lined with a carbonaceous film, giving the illusion of total preservation. Close studies of the husks nevertheless made it clear that the Oligocene ants were basically modern in form. All the species that lived in the Oligocene forests of Europe are now extinct, but 60 percent of the genera to which they belonged are alive today.

By Oligocene times, then, ants modern in outward appearance had come into full flower. Prior to 1966, myrmecologists had a clear picture of the Baltic amber and several other ancient faunas, yet they

still had no idea of the trunk and roots of the ants' family tree. Creationists had taken note of this absence in their campaign to discredit the theory of evolution. Ants, they argued, are an example of a group put on earth by a single act of special creation. Those of us reconstructing the evolutionary history of ants believed otherwise. We guessed that the earliest species were simply very scarce, and that the fossil beds containing them were just poorly explored, so that in time at least a few specimens would turn up. We believed that the missing link existed in deposits of early Eocene age, perhaps 60 million years old, or further back still, into the Mesozoic Era. The Ur-ant may well have stung an occasional dinosaur.

It would be nice to report, we wish it were true, that the crucial fossil was found in the headwaters of the Amazon by a brave graduate student who struggled downstream, malaria-racked and exhausted, to a remote mission village, his dugout canoe still pierced by broken arrow shafts. That he posted the specimen before proceeding on to Manaus for medical treatment and rest and to await congratulations from the jubilant research group at Harvard. But the truth is the Ur-ant was discovered by Mr. and Mrs. Edmund Frey, a retired couple living in Mountainside, New Jersey. They found it at the base of the seaside bluffs of Cliffwood Beach, a thickly settled middle-class residential area just south of Newark. The Freys sent an amber piece containing two worker ants to Donald Baird of Princeton University. Baird, recognizing its scientific importance, passed it on to Frank M. Carpenter of Harvard University, the world authority on insect paleontology and teacher of Edward Wilson.

Carpenter called Wilson on the telephone, two floors above him in Harvard's Biological Laboratories.

"The ants are here," said Carpenter.

"I'll be down in two milliseconds," Wilson replied, adrenalin surging.

Wilson ran down the stairs and into Carpenter's office, picked up the specimen, fumbled with it and dropped it on the floor, whereupon it broke into two pieces. Fortunately, each fragment contained an ant still in place and undamaged. Both pieces were composed of clear, pale golden matrix. When polished they provided beautiful views of the ants, wonderfully preserved, as though the insects had been entombed only the day before.

The amber was the fossilized resin of sequoia trees that grew at the Cliffwood Beach locality 90 million years ago, near the middle of the Cretaceous Period, when dinosaurs were the dominant large land vertebrates. The deposit in which they lay is a thin layer of light-colored sands bearing blackened lignitic fragments of sequoia wood. Scattered through the chunks of lignite are abundant tiny yellow grains of the resin. These fragments are usually all that can be found of the Cretaceous amber. But occasionally a larger piece turns up at Cliffwood Beach, and once in a great while it contains the remains of an insect. The Freys strolled along the beach soon after a storm washed away part of the bluff, exposing more of the fossilized wood. Alert to the possibility of finding amber, they had the extreme good luck to come upon the large piece containing the two ants.

Wilson put the fossils under the microscope and began to sketch and measure them from all sides. After several hours he picked up the telephone and called William L. Brown at Cornell University. Brown was a fellow specialist in ant classification who had for years shared his dream of finding a Mesozoic ant and thereby, perhaps, to learn the identity of the missing link to the ancestral wasps. Both men had guessed from comparisons of living species what traits the ancestral form might or, if evolutionary theory is correct, *should* possess. Wilson reported that the ants were indeed as primitive as expected. They had a mosaic of anatomical features found variously in modern ants or in wasps, as well as some that were intermediate between the two groups. The diagnosis of the Ur-ant was astounding: short jaws with only two teeth, like those of wasps; what appears to be the blisterlike cover of a metapleural gland, the secretory organ (located on the thorax, or mid-part of the body) that defines modern ants but is unknown in wasps; the first segment of the antennae elongated to give them the elbowed look characterizing ants, yet here, in the Mesozoic fossils, only to a degree intermediate between modern ants and wasps; the remaining, outer part of the antennae long and flexible, as in wasps; the thorax with a distinct scutum and scutellum (two plates forming part of the middle part of the body), also a trait of wasps; and an antlike waist, yet one that is simple in form, as though it had only recently evolved.

The ants of the New Jersey amber—we were bold enough to call them ants despite their mixed qualities—were about 5 millimeters long. We gave them the formal name *Sphecomyrma freyi.* The generic name *Sphecomyrma* means "wasp ant," and the species name *freyi* honors the couple who found and so quickly and generously donated them to science. In antic moments, noticing their well-developed stings, we imagined swarms of *Sphecomyrma* workers repelling small dinosaurs that brushed too close to their nest.

It had taken over a hundred years for entomologists studying insect fossils from all around the world to turn up these first Mesozoic ants. Then, abruptly, a great many more were found. Russian paleontologists, who are the most active in the study of ancient insect faunas, discovered specimens in Cretaceous deposits from three parts of the old Soviet Union: Magadan in northeastern Siberia on the shore of the Okhotsk Sea, the Taymyr Peninsula in extreme north-central Siberia, and Kazakhstan to the far south. Two more specimens were found about the same time by Canadian entomologists in Cretaceous amber from Alberta. When all the specimens are put together, they present the first crude picture of an ancient ant colony. Some are clearly workers, others queens, and still others male.

There is little variation among these companions of the dinosaurs. They can all arguably be placed in the single genus *Sphecomyrma,* a restriction in great contrast to the more than 300 genera, comprising many thousands of species, that make up the modern ant fauna. The Sphecomyrmas are also very scarce. They compose only about 1 percent of the insects in the Cretaceous deposits, in contrast to later assemblages, where ants are among the most abundant and diverse of the insects represented. The picture that emerged from the newly discovered fossils is one of a few rare species found all across Laurasia, the ancient supercontinent that combined present-day Europe, Asia, and North America. The continuity of this land mass allowed much easier dispersal than is possible today. The region in which the ants lived was probably warm temperate to subtropical. Far to the south, on the southern supercontinent Gondwanaland, comprising present-day Africa, Madagascar, South America, Australia, India, parts of South Asia, and Antarctica, ant evolution may have taken a different direction. A single specimen has recently been found by Brazilian paleontologists in Cretaceous rock deposits of Santana do Cariri in the eastern state of Ceará. The specimen, from 100 to 112 million years old, does not belong with *Sphecomyrma* but is much closer to the living, primitive bulldog ant of Australia. Described in 1991 by C. Roberto Brandão, formerly a student in the laboratory of Hölldobler and Wilson, it was given the name *Cariridris bipetiolata.*

Let us now proceed to Australia, where, about the time *Sphecomyrma* was discovered, a search of another kind was conducted for the most primitive *living* ant, as opposed to extinct species. Entomologists, it is clear, can learn a great deal about the evolution of anatomy and even the different castes of the earliest ants from fossils. But to piece together the history of social behavior they must study living forms. Their dream for generations has been that somewhere a species survives that still preserves the most primitive colonial organization, in other words composes a living fossil of behavior. Their hopes have centered mostly on Australia, home of other archaic life forms such as the egg-laying mammals, the platypus and echidna.

In the 1970s this dream was realized. The ant was *Nothomyrmecia macrops,* a large, yellow species with protruding black eyes and long mandibles shaped like the serrated blades of a dressmaker's pinking shears. For thirty-five years previously the species had been known to science from two museum specimens and nothing else. The primitive anatomy of *Nothomyrmecia*—vaguely wasplike appearance, with a simply constructed waist and symmetric, fine-toothed mandibles—was enticing. But to proceed to the next step, to rediscover the species and study living colonies, proved remarkably difficult and frustrating.

The long story began on December 7, 1931, when a small excursion party set out by truck from Balladonia Station, a sheep ranch in Western Australia, for a month's trip to the south through uninhabited eucalyptus scrub forest and sandplain heath. They proceeded for 110 miles past Mt. Ragged, a low, granitic hill at the western end of the Great Australian Bight, to the abandoned Thomas River farm. They then traveled west for

70 miles across the sandplain heath to the little coastal town of Esperance. This trek through a unique part of the Australian wilderness was taken mostly for pleasure. But the heath in which the party traveled is also one of the botanically richest places in the world, with large numbers of shrubs and herbaceous plants found nowhere else, and thus of great interest to biologists. Several people in the party had been asked to collect insects along the way. These they placed in jars of alcohol tied to the saddles of their horses, without recording the exact spot where they were found. The specimens, including two worker specimens of a large yellow ant, were then turned over to Mrs. A. E. Crocker, an artist living at Balladonia Station, who frequently painted specimens collected in this way. She eventually passed the insects on to the National Museum of Victoria, Melbourne, where the ants were described in 1934 by the myrmecologist John Clark as a new genus and species, *Nothomyrmecia macrops.*

William Brown, the doyen of myrmecology, was the first to recognize the evolutionary significance of *Nothomyrmecia.* He set out in November 1951 to collect more specimens, tracing part of the route of the 1931 expedition east of Esperance on the Thomas River trail. But with no firm idea of the collection site, and for reasons peculiar to the biology of *Nothomyrmecia* we will come to shortly, he failed. In January 1955, a second attempt was made by Wilson, joined by Caryl P. Haskins, then president of the Carnegie Institution of Washington and an ardent student of ant biology, and the celebrated Australian naturalist Vincent Serventy. Traveling by truck from Esperance along the 1931 route, they searched thoroughly around the Thomas River station and the sandplain heath north to Mt. Ragged, searching all the major habitats night and day for a week. No *Nothomyrmecia.*

By this time the "missing link" ants were becoming famous throughout Australia and among entomologists abroad—as famous as can be expected for an insect that does not carry malaria or destroy wheat crops. National pride was added to the mix, as other Australian entomologists and naturalists vied to be the first to rediscover *Nothomyrmecia* and study it in the living state in advance of their American rivals. All efforts failed, and aficianados began to speculate that either the locality had been wrongly recorded or the ant, like so many treasures of the Australian fauna and flora, had become extinct.

The break came, as advances often do in science, in a completely unexpected way. *Nothomyrmecia* was rediscovered, to the relief of local entomologists, by Robert Taylor, an Australian. After conducting his Ph.D. studies at Harvard under Wilson in the early 1960s, he joined the Division of Entomology at Australia's Commonwealth Scientific and Industrial Research Organization (CSIRO), located in the capital city of Canberra. In time he became chief curator of the Australian National Insect Collection. In that capacity he made it a personal mission to find the mystery ant.

In October of 1977, a spring month in Australia, Taylor led an expedition by truck from Canberra westward beyond South Australia. The group planned to drive along the Eyre Highway, which proceeds for a thousand miles across the barren Nullarbor Plain to the Mt. Ragged–Esperance area, specifically to search for *Nothomyrmecia.* They felt a sense of urgency, having learned that Bill Brown was about to undertake a do-or-die effort of his own. After proceeding 350 miles out of Adelaide, the group had vehicle trouble and was forced to stop and camp near the small town of Poochera. The site was surrounded by mallee, the multistemmed eucalyptus scrub that covers a large part of the semidesert regions of southern Australia. That night the temperature dropped into the fifties Fahrenheit (about 10° Celsius), and the entomologists huddled in warm clothing debating whether to collect insects that night. It seemed to be too cold for ants, let alone flying insects, to be active. In any case, *Nothomyrmecia* was assumed to occur a thousand miles west, more than halfway across the continent.

Bob Taylor, a questing and voluble scientist always on the outlook for ants, could not sit still that evening. He ventured into the mallee with a flashlight, just on the off chance that workers of some species or other might still be active in spite of the cold. A short time later he ran back into camp shouting, in the best Australian tradition, "The bloody bastard's here! I've got the *Notho*-bloody-*myrmecia!*"

He had discovered a *Nothomyrmecia macrops* worker crawling on a tree trunk, only twenty paces from the expedition camper-trucks. The secret of the ant had been revealed by the circumstances of the encounter. While it is true that *Nothomyrmecia* is both rare and restricted in distribution, so much so that the Red Data Book of the International Union for Conservation of Nature and Natural Resources (IUCN) now lists it as a possibly threatened species, it is also a cold-weather ant, one of those few species active when other ants, along with almost all entomologists, are indoors keeping warm.

In the years to follow, researchers descended on Poochera, elevating the hamlet to international fame (at least among entomologists). A large percentage of the world's ant specialists have stayed at the tiny hotel there. The Balladonia population of *Nothomyrmecia*, if it still exists sixty years after the first species were popped into a collecting jar, has not been entirely forgotten, but further efforts to find it, even in cold weather, have failed. More colonies may yet be discovered in the sandplain-mallee scrub or among the paperbark groves of the Thomas River station. Meanwhile field studies have proceeded to considerable detail at Poochera, and colonies have been removed for close laboratory analysis. Virtually every aspect of the life cycle and general biology of the species has been examined, and *Nothomyrmecia macrops* is now one of the most thoroughly understood of all ants.

What we have learned about this species can be summarized by saying that it has a very simple social organization, as expected. In particular, the queens are closely similar in appearance to the workers. No worker subcastes exist, such as soldiers specialized to defend the nest, and each worker appears to perform the same tasks. The colonies are small, with populations never exceeding a hundred adults. Eggs laid by the queen are left scattered singly on the nest floor, not stacked into piles in the manner of most higher ants. Like wasps, the workers gather two kinds of food: nectar to feed themselves and insect prey mainly to give the larvae.

Little contact occurs among adult *Nothomyrmecia*. They do not pass regurgitated food back and forth to one another, as do most higher ants. The queens, normally at the center of attention in other ant colonies, are mostly ignored. The workers forage in a solitary manner, and when they discover food outside the nest they bring it home alone, without trying to recruit their nestmates. They attack and sting flies, hemipteran bugs, and a variety of other insect prey. So far as we know workers use only two forms of chemical communication: alarming nestmates when enemies are discovered and distinguishing nestmates from alien *Nothomyrmecia* by their shared body odor.

The nests of this archaic ant are simple chambers dug in the soil and connected by tunnels. The life cycle also has a generalized quality. Virgin queens leave their home nest to be mated, dig a nest on their own, and then, wasplike, forage away from the nest to collect food. Like queens of paper wasps and other primitively social wasps, several young *Nothomyrmecia* queens sometimes cooperate to dig a nest and raise a first crop of workers. Later, however, one dominates her partners by standing above them at frequent intervals, and eventually the first crop of workers expels the losers altogether by dragging them from the nest. So colonies found at Poochera, most already well established when excavated, have always had only a single mother queen. The affinity of the workers for low temperatures is an odd trait, but one that may represent no more than an adaptation of the species to life in the cold temperate zone of Australia.

In the consistent simplicity of their colonial organization, the Nothomyrmecias are close to what is reasonably supposed to be the evolutionary level reached by the first social Mesozoic ants. They possess a few of the intimate habits of more advanced ant species, including the tendency to groom the bodies of their nestmates. But in most regards their behavior is close to what we expect from solitary wasps that invented cooperative sisterhoods, changed their anatomy a bit, and became the first ants. The societies of ants arose, it seems, from aggregates of Mesozoic solitary wasps that were already building nests in the soil and retrieving insect prey to feed their larvae, just as many solitary wasps do today. The essential first step in the process was for the mother to stay with her young after they became adults. All that was required afterward in order to achieve colonial life was for the daughters to curtail their personal reproduction and to assist the mother in raising more sisters.

Two other ants with primitive anatomy are known to possess similar elementary social habits. They are the Australian bulldog ants of the genus *Myrmecia,* which are similar in appearance to *Nothomyrmecia;* and *Amblyopone,* an evolutionarily very distinct group that is worldwide in distribution but most abundant and diverse in Australia. Until *Nothomyrmecia* was rediscovered, *Myrmecia* provided the paradigm for "primitive" ant social organization. Its behavior is now known to be substantially more advanced than that of *Nothomyrmecia.*

Our guess is that the most anatomically wasplike ants of all discovered to date, *Sphecomyrma,* behaved much like *Nothomyrmecia* and the other living primitive ants. But we will never know for sure. Because there are no known solitary ants, possessing the basic anatomy of ant queens but living alone or in small groups without workers, we are unlikely to succeed in digging much deeper into the roots of social evolution. Yet barring surprises, in science always a possibility, we believe the story we and other entomologists have pieced together is close to what actually happened more than a hundred million years ago.

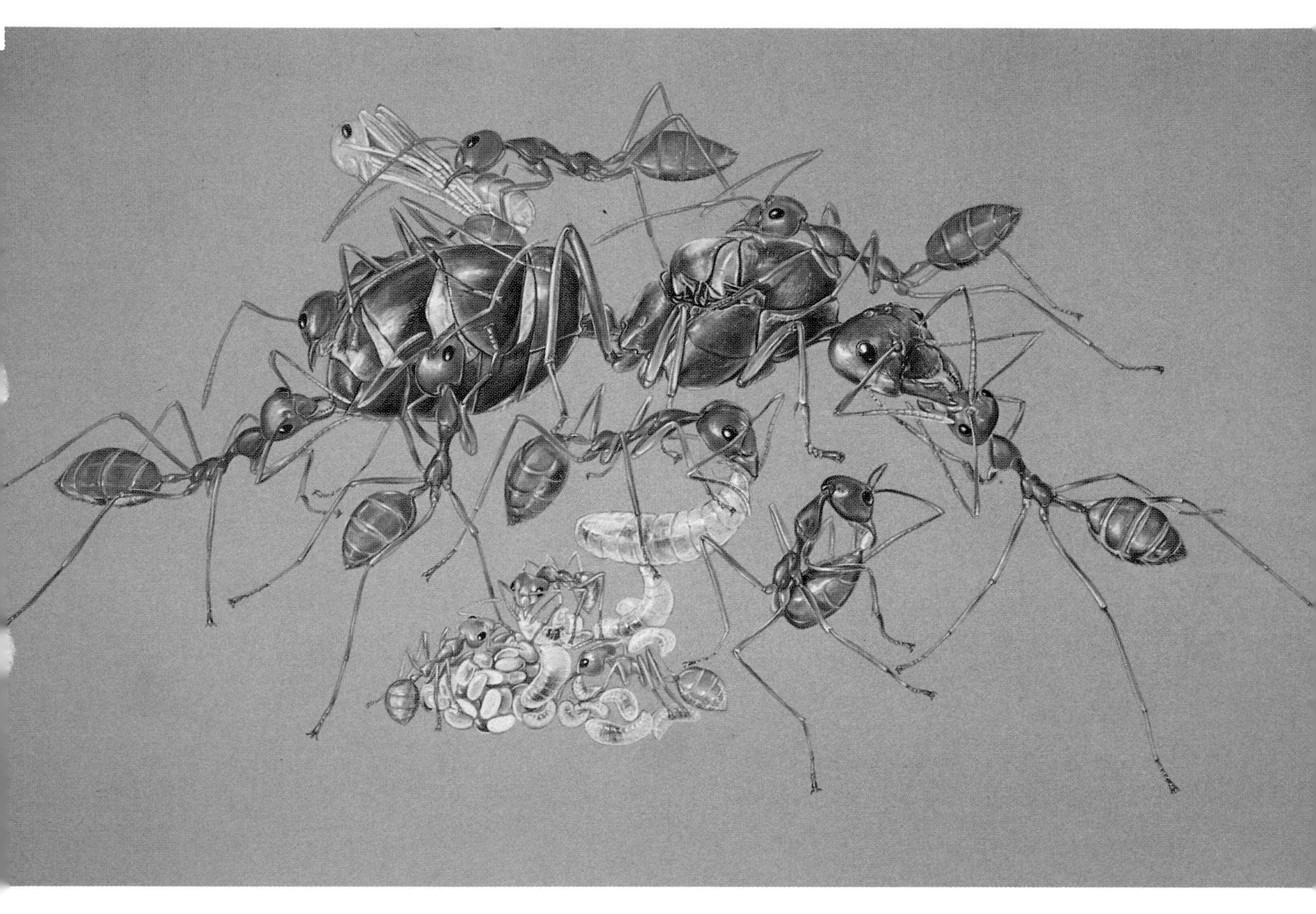

A mature colony of weaver ants contains a single mother queen and more than half a million workers. The fragment shown here includes the queen. She is groomed continually by major workers, which also forage, build and defend the nest, and care for the large larvae. In the foreground a group of minor workers attend to one of their principal tasks, the care of the eggs and small larvae. (Painting by Turid Forsyth.)

Social carrying in ants is a common means of recruiting nestmates to new nest sites. If the transported ant is "satisfied," it too may return to the old nest and start to carry nestmates on its own. Shown here are carpenter ants of Australia, *Camponotus perthiana.*

Facing page
Social carrying in African weaver ants *(above)* and ponerine ants, *Ectatomma ruidum*, of tropical America (*below*).

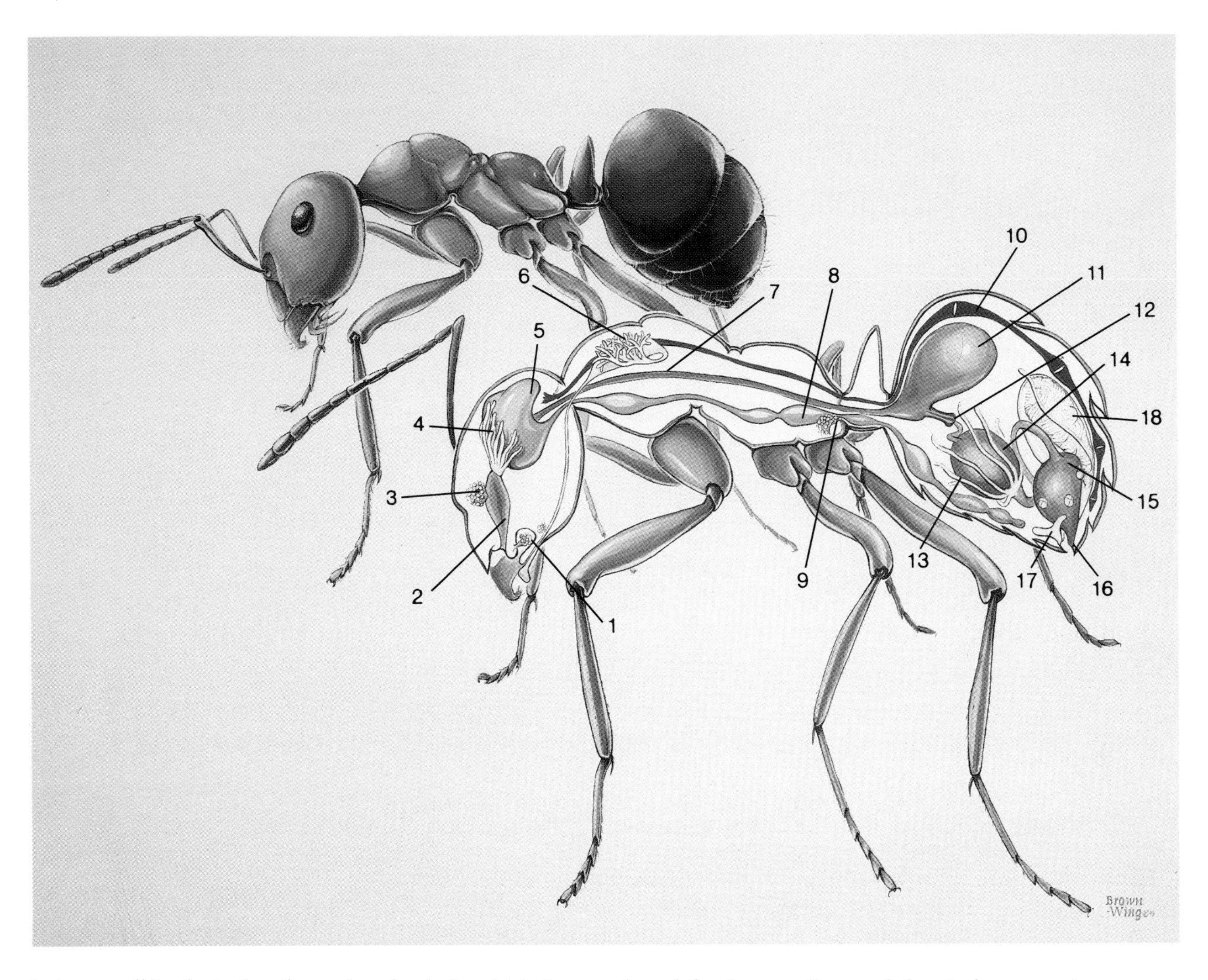

Ants are walking batteries of exocrine glands, in which they produce defensive secretions and chemical communication signals. One such system of glands is shown here as part of the internal anatomy of a *Formica* ant. The brain and nervous system are shown in blue, the digestive system in pink, the heart in red, and the glands and associated structures in yellow: *(1)* mandibular gland; *(2)* pharynx; *(3)* propharyngeal gland; *(4)* postpharyngeal gland; *(5)* brain; *(6)* labial gland; *(7)* esophagus; *(8)* nervous system; *(9)* metapleural gland; *(10)* heart; *(11)* crop; *(12)* proventriculus; *(13)* Malpighian vesicles; *(14)* midgut; *(15)* rectum; *(16)* anus; *(17)* Dufour's gland; *(18)* poison gland and reservoir. (Painting by Katherine Brown-Wing.)

Facing page
A nest of the honeypot ant *Myrmecocystus mimicus* of the southwestern United States. Repletes, which hang from the nest ceiling, are large workers whose crops are grossly distended by liquid food. They regurgitate part of the liquid back to their nestmates during times of food shortage; one such exchange is occurring between two workers in the foreground. The queen can be seen beyond the pile of cocoons and larvae. (Painting by John D. Dawson, courtesy of the National Geographic Society.)

J. DAWSON

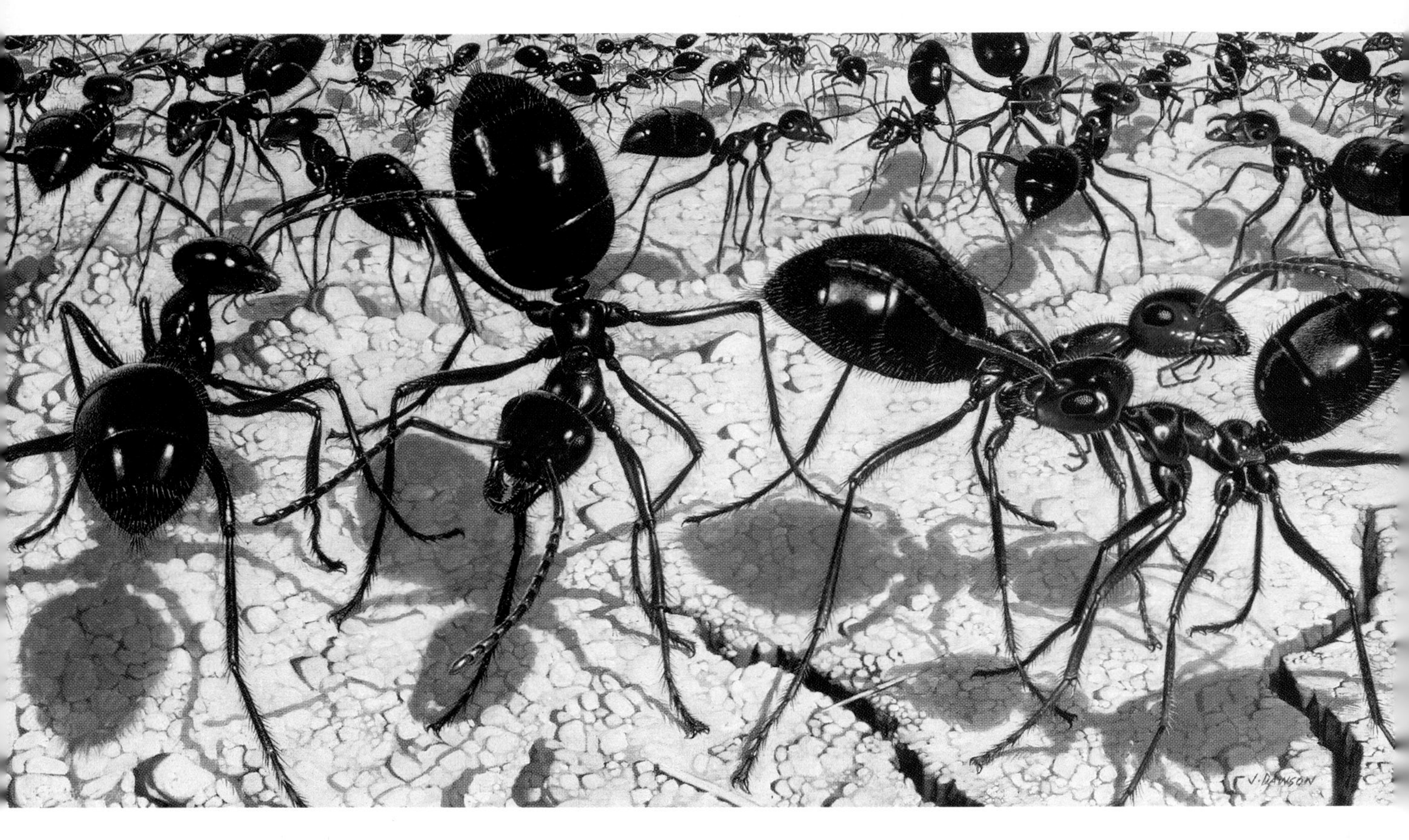

Territorial tournaments of *Myrmecocystus mimicus.* Workers from neighboring colonies confront one another in ritual displays during which they raise their heads and abdomens while walking high with their legs in stiltlike position. (Painting by John D. Dawson, courtesy of the National Geographic Society.)

Facing page
During territorial tournaments, workers of opposing honeypot colonies meet face on, then try to dislodge each other by pressing sideways *(above)*. They also try to stand taller (*below*). During such exchanges, the ants evidently assess the size of their adversaries, an important factor in the outcome of the tournaments and occasional wars.

Territorial displays between honeypot ant colonies can on occasion give way to raids, during which the stronger colony kills the queen of the weaker colony, pillages the brood *(above)*, and kidnaps the repletes *(below)*.

During a raid the victorious honeypot raiders are sometimes robbed of their booty—in this case a replete worker from the defeated colony—by workers of another species, *Forelius pruinosus.* Although only a fraction of the size of the honeypots, the *Forelius* are able to defeat their victims by force of numbers and the use of powerful chemical sprays.

Forelius raiders using a chemical spray to drive honeypot ants back into their nests. The spray, a kind of ant mace, is emitted from the pygidial gland at the tip of the abdomen.

In the Arizona desert, workers of *Conomyrma bicolor* inhibit foraging of *Myrmecocystus mexicanus* by dropping pebbles down the nest entrance of the *Myrmecocystus.* (Painting by Katherine Brown-Wing.)

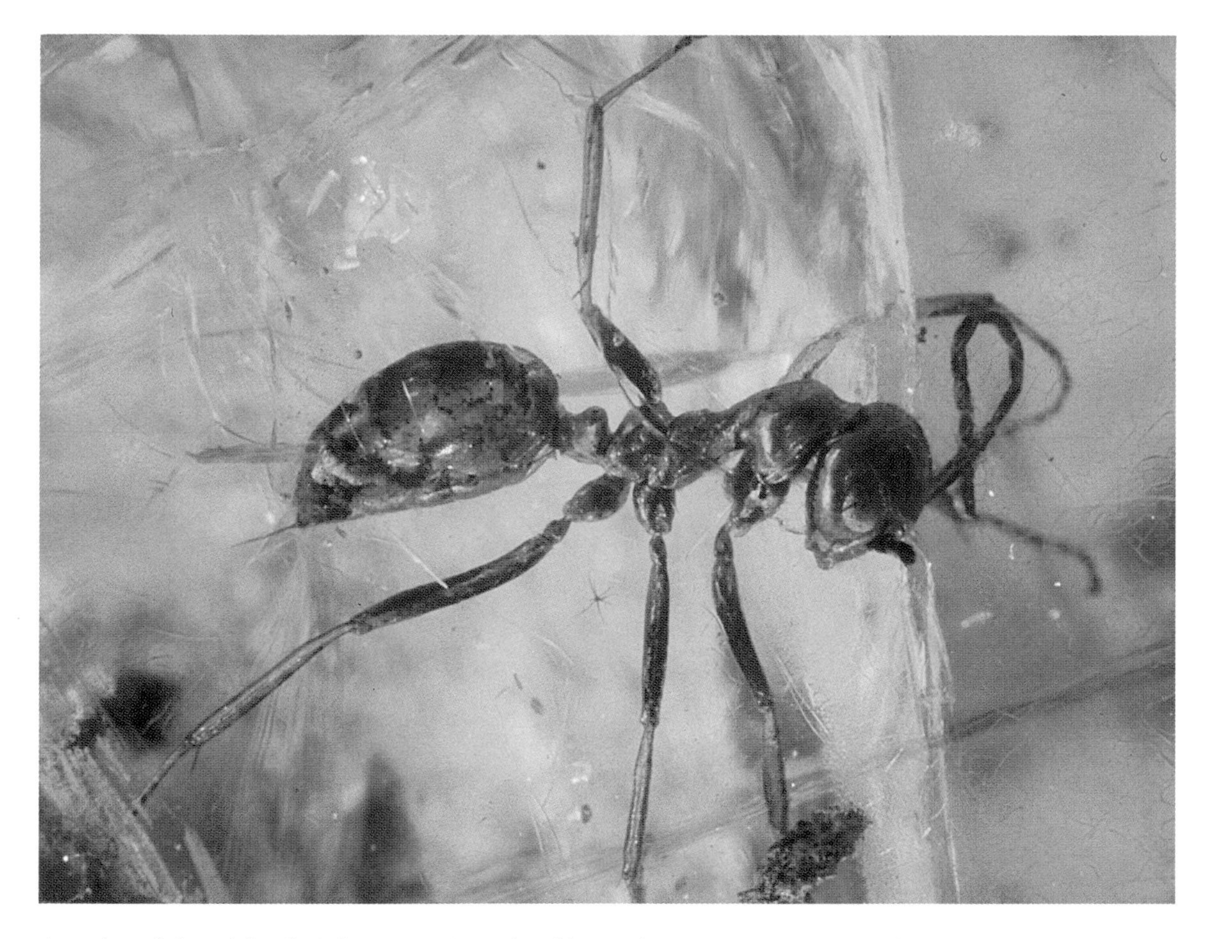

A worker of the subfamily Sphecomyrminae, the oldest and most primitive group of ants. This worker, the first described specimen of *Sphecomyrma freyi* and the entire ant subfamily Sphecomyrminae, was found in sequoia amber from New Jersey. It dates from the lower part of the Upper Cretaceous and is approximately 80 million years old. (Photograph by Frank M. Carpenter.)

Facing page
Competition through robbing. *Above:* a honeypot ant *(Myrmecocystus mimicus)* (marked with a blue dot) robs a termite prey from a *Pogonomyrmex* forager in the Arizona desert. Such workers often lie in wait close to the nest entrance of *Pogonomyrmex. Below:* when attacked the honeypot ant runs away swiftly, circles back, and quickly repositions itself near the entrance.

Part of the colony of the primitive Australian ant *Nothomyrmecia macrops,* with the queen in the foreground; also present are workers, larvae, pupal cocoons, and winged males. (Photograph by Robert W. Taylor, CSIRO, Canberra, Australia.)

Facing page
At the top, a *Nothomyrmecia* huntress carrys a captured wasp back to her nest. This photograph was taken near Poochera in south central Australia, site of the only known population of this primitive ant. The picture at the bottom shows the habitat where *Nothomyrmecia* was found.

Besides *Nothomyrmecia macrops,* two other ant species considered phylogenetically primitive are the bulldog ant *Myrmecia pilosula (above)* and *Amblyopone australis (below),* both from Australia.

Conflict and Dominance

IN 1950, WILSON, then a 20-year-old student at the University of Alabama, addressed an important problem in the study of fire ants. The imported species, which was in the process of spreading through the southernmost United States after its introduction from South America, existed in two color phases, one red and the other dark brown. It is now known that the two forms are actually full species in their own right. The red species is properly called *Solenopsis invicta* and the dark brown one *Solenopsis richteri.* The two hybridize freely in the United States, but only to a limited degree in South America. Each is distinguished not only by color but by a unique combination of traits in anatomy and biochemistry. At this early stage in the research, in 1950, it was important to determine whether the color difference is based on genes or whether it is merely the result of different living conditions.

Fire ants cannot be bred easily in the laboratory, like fruit flies, to search for the existence of color genes. The conditions required for mating are too precise, the life cycle too long and complicated. But the influence of genes versus environment could be tested indirectly, Wilson reasoned, by rearing the young of red queens with brown workers, or brown queens with red workers, to see if the color changed in the next generation to match that of the workers. If change did not occur, and if all the other conditions in the laboratory nests were kept the same as for control colonies (in which queens and workers were the same color), the environmental hypothesis would be eliminated and the genetic hypothesis upheld.

The adoption of one color form by another did prove feasible. Wilson discovered that he could induce fire-ant workers to tolerate alien queens if he first removed their mother queen and then chilled them to immobility and placed the alien among them. When the ants warmed up and became active again, they accepted the alien queen and reared the eggs she laid.

The color remained the same in the experiments, and the genetic hypothesis was upheld. The existence of color genes was not proved beyond reasonable doubt by this means, but at least it was strongly indicated. Then a strange thing happened. Wilson decided to play

around with his adoption technique, introducing up to five queens instead of one, just to see what would happen. All these attempts succeeded in full, but only for a while. After a day or two the workers began to execute the surplus queens by spread-eagling them and stinging them to death. They continued until only one was left. The winner was then kept and nurtured by the workers as their full-fledged nest queen. The workers never made a mistake. Not once did they go too far and kill the final survivor, thus dooming the colony as a whole.

The fire-ant adoption study was one of the first to indicate that not all is peace and harmony within ant colonies, even in species with a high degree of organization. Somehow the multiple queens were competing among themselves for the workers' favor in a life-and-death contest. As the years passed, more evidence accumulated to reveal that conflict and dominance among nestmates are widespread in the ants. Even more interesting, this strife in the sisterhood often goes far beyond mere squabbling. In many species it has been strongly ritualized during evolution, and has come to serve a prominent role in the regulation of the colony life cycle.

One striking example of the connection was revealed when Hölldobler and his student Stephen Bartz conducted a close study of colony founding in *Myrmecocystus mimicus*. This is the species of large honeypot ants, abundant in the deserts of Arizona and New Mexico, on which Hölldobler had based his studies of colonial warfare and "diplomacy." Each July, after the first summer rains have softened the hard, desiccated soil, the queens and males emerge in large numbers to conduct their nuptial flights. The queens, after being inseminated, descend to the ground, shed their wings, and dig burrows to start their own colonies. When Hölldobler excavated a large number of these founding nests—easy to do using nothing more than a trowel—he found that most were occupied by more than one queen.

By this time, the late 1970s, the alliance of multiple queens during colony founding was known to occur in many kinds of ants. Entomologists had even coined a special term for it: pleometrosis. But the alliances were also known to be short-lived. They seldom led to polygyny, the permanent or at least long-term association of queens in an older colony. Either the workers eliminate the excess queens, the procedure followed by fire ants, or the queens fight among themselves, sometimes assisted by aggressive workers who take the side of one or another.

This whole procedure might seem at first not to make good Darwinian sense. Why should a queen cooperate if she stands a good chance of being killed for it? One key advantage, revealed by additional studies in the 1960s and 1970s, is that multiple queens raise larger first broods of workers and in a shorter period of time than solitary queens. Thus colonies founded by multiple queens get a fast start when it is most needed. They are able to defend against enemies more quickly and to establish territories with greater efficiency soon after leaving their mother's nest to search for food. For a cooperating queen, this advantage is evidently great enough to outweigh the risk of an early death.

Walter Tschinkel, a researcher at Florida State University, observed that a larger fighting force is decisive in wars among fire-ant colonies, which are conducted frequently and at high intensity. A young colony unable to defend itself against its neighbors is quickly eliminated. Bartz and Hölldobler, in an independent study, discovered that the same phenomenon occurs in honeypot ants. When workers first emerge onto the desert floor, they start attacking other beginning colonies in the neighborhood as soon as they can find them. If victorious, they transport the brood to their own nests. The colony that wins a contest in the first encounter thus has an immediately larger force and an advantage over its remaining competitors who have not conducted a successful raid. Victory is piled upon victory until finally all the brood in the near vicinity ends up in one nest. In the process workers often abandon their own mothers in favor of the winning raiders, the ant equivalent of "better red than dead." In 23 such rounds in clusters of laboratory colonies watched by Bartz and Hölldobler, winners were always colonies that had been founded by multiple queens. In 19 of the cases they were the colonies with the largest number of queens of any colony in the neighborhood.

Once a honeypot colony has acquired a worker force large enough to secure itself against its neighbors, a new struggle commences, this time among the queens. In a typical exchange, one queen stands above her rival, occasionally stepping on her, while pointing downward with her head. The subordinate crouches low and holds still. Any queen that consistently yields to others is eventually driven from the nest by the workers, even though some of the aggressors are likely to be her own daughters.

Dominance struggles for reproductive rights also occur among queens in older, mature colonies. Jürgen Heinze, an associate of Bert Hölldobler in Würzburg, discovered the phenomenon in several species of *Leptothorax.* Dominance rituals among the queens lead to the establishment of functional monogyny, in which only the queen at the summit of the social hierarchy reproduces. The Brazilian entomologist Paulo Oliveira and his collaborators found these rituals to be frequent in the large tropical American hunting ant *Odontomachus chelifer,* in which several egg-laying queens commonly live close together. When a subordinate is challenged by a higher-ranking queen, she crouches, shuts her long, powerful mandibles, and pulls her antennae back out of reach. If she tries to raise her body, the dominant ant seizes her by the head. If she struggles in an attempt to free herself, the dominant may lift her bodily off the ground. She then gives up altogether, by pulling her legs in to her body in the "pupal posture" by which ants allow themselves to be carried from one place to another.

The queens of some ant species use a more subtle method of control. They do not challenge their rivals to combat, but instead pull their eggs from the brood pile and eat them. Those who destroy the largest number of opponents' eggs while losing the fewest of their own are in effect dominant, at least by Darwinian standards: their daughters will be disproportionately represented among the workers and in the next generation of queens.

The primary nest queens of other ant species are even more subtle. They produce inhibitory pheromones, chemical substances that prevent the production of eggs in the ovaries of virgin queens and workers. If the weaver-ant queen is removed, some of the workers will begin to lay eggs. But if the queen dies and her corpse is left in the nest, so that she continues to exude the pheromone even after death, the workers will remain infertile.

The more carefully entomologists have examined the fine details of

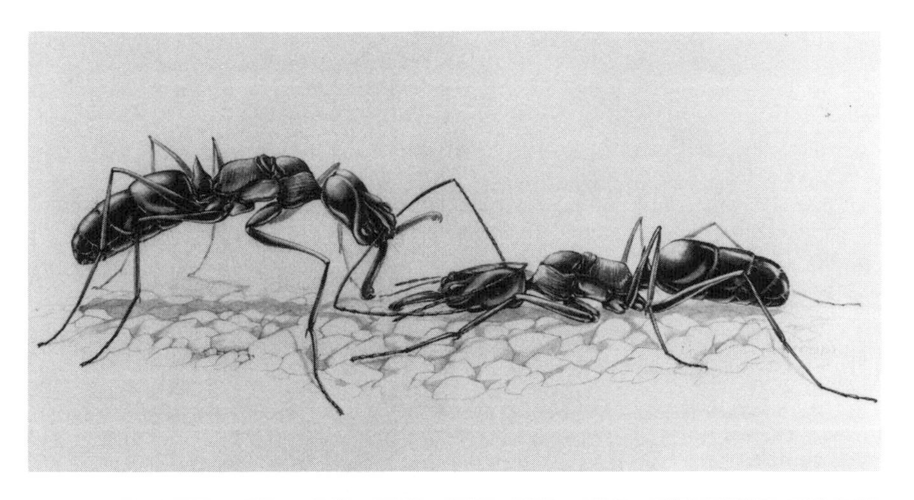

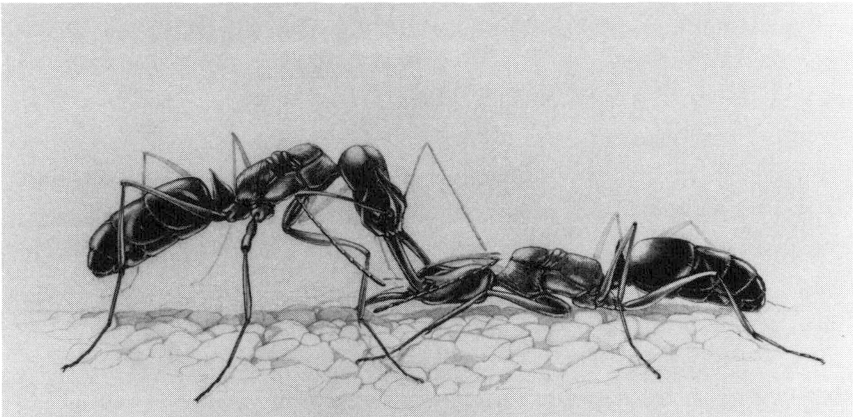

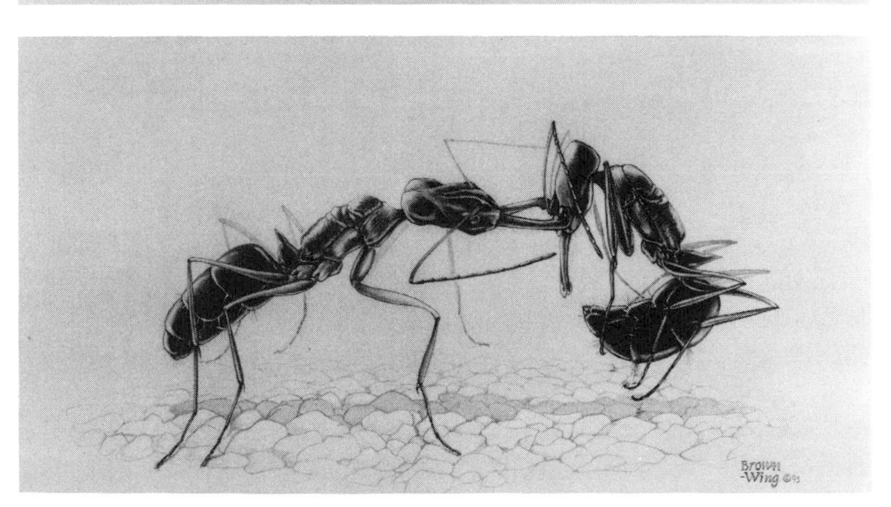

Dominance behavior between nestmate queens in the tropical American predatory ant *Odontomachus chelifer.* At the top the dominant queen threatens with opened mandibles, while her sister crouches in submission. In the center the dominant escalates the conflict, grasping the subordinate by the head and then, at the bottom, lifting her off the ground. When handled this way, the sister ant indicates submission by pulling her legs into the immature pupal position. (Drawings by Katherine Brown-Wing.)

colony organization, the more extensive and complex have been the conflicts revealed. To pay close attention to the relationships of particular individuals is like moving into an outwardly peaceful city only to realize, after living there for a while, that the place is rife with family quarrels, theft, street muggings, and even murder. Dominance struggles occur even among ant workers of the same colony. Blaine Cole, an American entomologist, first demonstrated the phenomenon conclusively by marking workers of the Floridian species *Leptothorax allardycei* so that they could be followed individually. Conflict, he observed, reaches a peak when the mother queen is removed. He estimated that the most competitive workers in queenless colonies spend more time threatening and pummeling one another than they do taking care of the brood. The *Leptothorax* workers are so self-serving that even when the queen is present the most dominant individuals lay 20 percent of the eggs. Such eggs are unfertilized and therefore destined, should they survive at all, to produce males. High-ranking workers also consistently receive more food, which allows them to grow large ovaries filled with eggs.

Conflict has become extremely ritualized in many of the ponerine hunting ants, most of which live in remote tropical localities and hence have only recently been studied. Christian Peeters, a myrmecologist working in Paris, has devoted most of his research to ritual conflict and other behaviors that affect reproduction in ponerines. With a Japanese colleague, Seigo Higashi, he found one of the most surprising cases in the Australian species *Diacamma australe.* These large, swift-running ants have no queens. All the females, which are anatomically workers, emerge from the cocoon bearing tiny, budlike vestigial wings called gemmae. The most dominant worker, who lays the eggs, bites off the gemmae of her nestmates soon after they appear. The mutilation inhibits the development of their ovaries, consigning them permanently to worker status. Only the dominant worker mates with males, and only she reproduces. When her gemmae are surgically removed in the laboratory, however, she becomes timid and is transformed into a functional worker.

As bizarre as the *Diacamma* dominance system may seem, it is outdone by the system that Christian Peeters and Bert Hölldobler recently discovered in the Indian ponerine ant *Harpegnathos saltator.* Large colonies of this species are class-ridden societies tense with the maneuvers of shifting status. The interactions among colony members bear a remarkable albeit superficial resemblance to some forms of human political behavior.

New colonies of *Harpegnathos* ants are evidently founded in a conventional manner by inseminated queens. As the colony grows, however, the queen disappears and a group of mated, fully reproducing workers, called gamergates, takes over. The colony history, then, proceeds in three stages. Small colonies, in the early period of growth, comprise a reproductive queen and a few sterile workers. Middle-sized colonies still have a queen but in addition they have mated and unmated workers. Finally, large colonies, containing about 300 or more adult members, lack a queen and are composed instead entirely of mated and unmated workers.

As a consequence of this life cycle, the membership of a large *Harpegnathos* colony is organized into three social classes. On top are the dominants, which possess fully developed ovaries and lay all the eggs. On the bottom is the class of virgin subordinates. Some of them are destined to move into the top class; about the time of their rise in rank they mate with visiting males and become reproductives. Others, however, remain in the lowest class and serve as nurses, nest builders, and foragers throughout their lives. Finally, the third class consists of mated subordinates. They in turn are of two kinds: workers who managed to mate even without rising in status, and formerly dominant gamergates who have been displaced downward in rank by more competitive nestmates. The individual fates of members of the third class are dependent on the future health and behavior of their rivals.

Status in this complex class system is settled by a ritualized form of dueling, in which the workers wield their antennae like whips. The exchanges begin when one ant lashes the other, then thrusts her body forward, causing the opponent to walk backward. After the pair have traveled about a body length, with the first ant continuing to lash the other, the entire process is reversed. The second ant now forces the first ant to retreat. As an aggressor attacks, she pulls the first, long segment of each antenna back against the side of her head while extending the more flexible outer segments of her antennae toward the opponent. The force

Ritual dueling among nestmates of the Asian ant *Harpegnathos saltator.* In the typical sequence depicted here (from top to bottom), the advancing worker lashes the backward moving sister ant with her antennae. After the pair have traveled in this manner for about a body length, the procedure is reversed, and the lashed becomes the lasher. (Drawings by Malu Obermayer.)

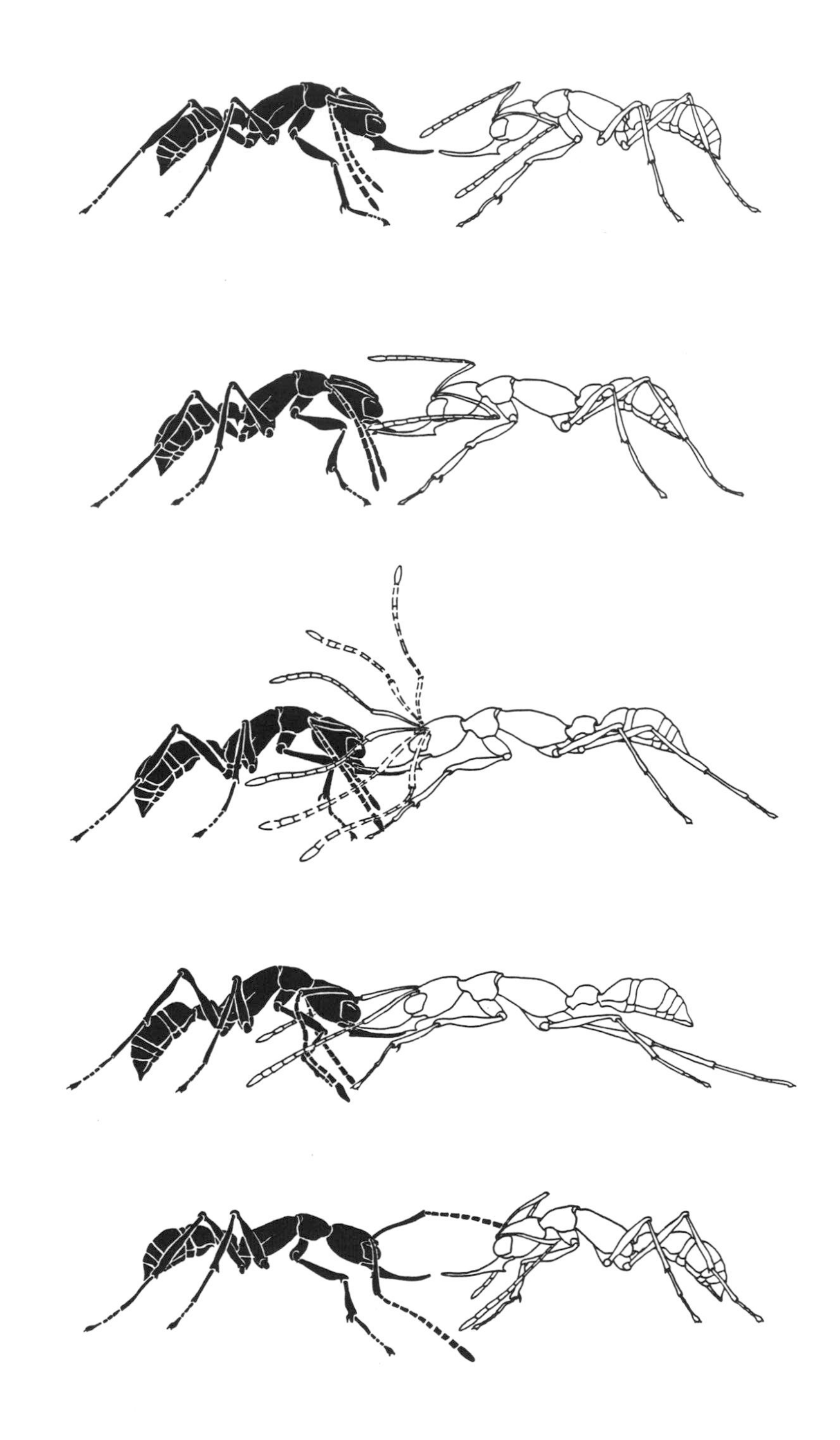

of each lashing strike on the opponent's body is strong enough to bend the outer segments back at the point of impact.

After this odd pas de deux is repeated up to 24 times, the two combatants simply walk away from each other. There is no obvious winner, and the whole performance appears to have been no more than a reaffirmation of social equality. The *Harpegnathos* sometimes use a second form of combat, however, that is more decisive. One ant approaches another while whipping her antennae from side to side. The maneuver is used most frequently by subordinates in an effort to overcome a higher-ranking nestmate. Often the challenged worker simply ignores the approach and continues any activity in which she is engaged at the moment. Alternatively, she may initiate a back-and-forth dueling match. Or she may escalate the exchange by launching an all-out assault, using her long mandibles to seize and thrust the opponent violently downward.

The *Harpegnathos* nestmates are often much more chaste in their encounters, displaying restraint and a certain politesse. As a dominant approaches, a subordinate may simply lower her body and draw her antennae back. The dominant may respond by stepping on the head of her nestmate, while nipping lightly at her body. The two then part amicably.

Life in the *Harpegnathos* colony is not always filled with conflict. There are periods of apparent complete tranquillity, in which no dominance interactions are to be observed. The peace is eventually broken, however, as one worker or another rising in rank from the lower classes decides to challenge a member of the top class. Her action precipitates a frenzy of ritual dueling among the dominant workers, who scurry about as if to cement their own high status among their peers. At the same time these workers also attack any nestmates from among the lower ranks who dare challenge them. In this endeavor they do not always prevail. Some are demoted to the mated middle class and their places are taken by former subordinates. And so the society moves along in Heraclitean progression, outwardly always the same but inwardly forever changing.

The Origin of Cooperation

MOST OF BIOLOGY comes down to two kinds of questions: how things work, and why they work. To put it another way, how is a process accomplished by anatomical and molecular actions, and why in the course of evolution did it come out that way and not some other? Biologists think they know basically how ant societies work and the approximate time they came into being: 100 to 120 million years ago. The time has come to ask *why* this important event occurred. What was the advantage of social life hit upon by the ancestral wasps that turned them into ants?

The single most important quality of the ant colony is the existence of the worker caste, which comprises females subservient to the needs of their mother, content to surrender their own reproduction in order to raise sisters and brothers. Their instincts cause them not only to give up having offspring on their own but also to risk their lives on behalf of the colony. Just leaving the nest to search for food is to choose danger over safety. Researchers have found that when harvesting ants of a western United States species *(Pogonomyrmex californicus)* forage, they suffer a death rate of 6 percent per hour due to fighting with neighboring colonies. Still other workers die from attacks by predators or lose their way. This casualty rate is high but not unique. Virtual suicide is also the fate of workers of *Cataglyphis bicolor,* a scavenger of dead insects and other arthropods in the North African desert. The Swiss entomologists Paul Schmid-Hempel and Rüdiger Wehner discovered that at any given time about 15 percent of the workers are engaged in long, dangerous searches away from the nest, at which time they are preyed on heavily by spiders and robber flies. On average each forager lasts only a week, but in that short interval she manages to collect 15 to 20 times her own body weight in food.

Why then (to return to the second great question of biology) do ants behave in such an altruistic manner? First consider the larger question of the origin of *any* kind of social behavior. What is the Darwinian advantage of living in a group? The correct answer is also the most obvious one. If an animal survives more consistently and

has more offspring across its lifetime as a member of a group, then it is better off cooperating than continuing as a solitary. The evidence shows that such is indeed generally the case in nature. Birds in flocks and elephants in herds, for example, do live longer and have more offspring than when they live alone. Because of the power of their group, they find food more quickly and defend against enemies with a greater expectation of victory.

The hypothesis of strength through numbers works best for simple animal societies, whose members cooperate but still look out for their own personal interests. It is not enough, however, to explain the amazing sacrificial nature of the ant workers. These selfless females die young and they seldom leave offspring.

The puzzle of ant altruism has played a historic role in the study of animal behavior. For generations biologists have attempted to fit the phenomenon to the Darwinian theory of evolution by natural selection. In so doing they often resort to complex explanations. The prevailing theory as we write is evolution by kin selection, a modified form of natural selection which, like the original version of the theory itself, was first conceived by Darwin. Kin selection is the favoring or disfavoring of certain genes in relatives by actions taken on the part of an individual. Suppose, for example, that a member of a family chooses to be celibate and have no children, while nevertheless devoting herself to the welfare of her sisters. If the sacrifice causes the sisters to bear and raise more children than would otherwise be the case, the genes shared by the spinster and her siblings will be favored in natural selection and spread more quickly through the population. Sisters of ordinary animals (and human beings) on average share half of their genes by common descent. Put another way, half their genes are identical by virtue of their being born from the same parents. All the altruist has to do is more than double the numbers of children raised by one sister in order to make up for the genes she will lose in future generations by not having children of her own. That in essence is kin selection. If in addition some of the genes spread by this means predispose individuals to altruistic behavior, the trait can become a general characteristic of the species.

This idea was stated in very general form, without calculating numbers of genes, by Charles Darwin in *On the Origin of Species.* Darwin had a strong interest in ants and other social insects. He watched them around his country house at Downs, close to London, and visited the British Museum of Natural History to learn more about them from the entomologist Frederick Smith. He found in ants the "one special difficulty, which at first appeared to me insuperable, and actually fatal to my whole theory." How, the great naturalist asked, could the worker castes of insect societies have evolved if they are sterile and leave no offspring?

To save his theory, Darwin introduced the idea of natural selection operating at the level of the whole family rather than that of the single organism. If some of the individuals of the family are sterile, he reasoned, and yet important to the welfare of fertile relatives, as is the case for insect colonies, selection at the family level is not just possible but inevitable. With the entire family serving as the unit of selection, in the sense that it struggles against other families for survival and reproduction, the capacity to create sterile but altruistic relatives is favored during genetic evolution. "Thus a well-flavoured vegetable is cooked," he wrote, "and the individual is destroyed; but the horticulturist sows seeds of the same stock, and confidently expects to get nearly the same variety; breeders of cattle wish the flesh and fat to be well marbled together; the animal has been slaughtered, but the breeder goes with confidence to the same family." So sterile worker castes could be produced and sacrificed by ant colonies like an apple harvested from a tree or a steer selected and butchered from a herd of cattle, and still their genes would flourish in the surviving relatives. Speaking of the soldiers and minor workers of an ant colony, Darwin continued, "With these facts before me, I believe that natural selection, by acting on the fertile parents could form a species which regularly produce neuters, either all of a large size with one form of jaw, or all of small size with jaws having a widely different structure; or lastly, and this is the climax of our difficulty, one set of workers of one size and structure, and simultaneously another set of workers of a different size and structure."

Darwin had defined the principle of kin selection in elementary fashion to explain how self-sacrifice can arise by natural selection. Perhaps more to the point, he showed how ant workers can be removed as an impediment to his theory. He laid this key objection to rest. For one hundred years entomologists slumbered in the knowledge that sterile castes created no great theoretical problem of any kind. Why do insect societies arise? They assumed it was because of the advantages of communal life, and sterile castes seemed just a logical extension of the process. There seemed no need to explore the matter further.

Then, in 1963, the British entomologist and geneticist William D. Hamilton added a twist that reopened the subject in a startling manner. He said, in brief, that the Hymenoptera, the order of insects comprising bees, wasps, and ants, are genetically predisposed to become social because of the way they inherit sex. Kin selection works as Darwin said all right, but because of the quirky way sex is determined in the Hymenoptera, it is turned into a driving force. To see how this works, first consider the general quantitative principle of kin selection established by Hamilton. He said that in order for an altruistic trait to evolve, the benefit to relatives must outweigh the inverse of the degree of relationship between the donor and the relatives. Take the case in which the donor gives up her life, or at least remains childless, in order to help a relative. An individual ordinarily shares half her genes with a brother or sister; the inverse of one-half is two; the self-sacrifice must therefore more than double the offspring of the brother or sister if the gene for altruism is to increase in the population. The altruist also shares one-fourth her genes with an uncle; if her sacrifice is spent in that direction she must increase reproduction in the uncle more than four times for the gene to spread. To continue, she shares one-eighth her genes with a first cousin; the cousin's reproductive success must be boosted more than eight times for the gene to spread. And so on. The benefits can be bestowed this way cumulatively among many relatives. But outside the tight circle of immediate relatives, bounded by immediate descendants, and first cousins, the degree of relatedness falls off so steeply as to be difficult to detect. True altruism—instinctive generosity and sacrifice without expectation of personal repayment—is likely to exist only among members of the immediate family. Hereditary altruism, in short, is narrowly focused.

Now we come to Hamilton's twist for the hymenopterans. Members of the insect order Hymenoptera, comprising the ants, bees, and wasps, inherit sex by haplodiploidy. Despite the technical-sounding name, the procedure is the simplest known: fertilized eggs, which are diploid (possessing two sets of chromosomes), become females; unfertilized eggs, which are haploid (one set of chromosomes), become males. Hamilton noticed that because female hymenopterans have both a mother and a father, each contributing an equal number of genes, mothers share one-half their genes with their daughters. This is the usual circumstance in the animal kingdom. But sisters share *three-fourths* of their genes. This exceptionally close relationship is due to the fact that their father came from an unfertilized egg. Therefore he doesn't have a mix of genes, the usual condition, but instead carries just one set, which he got from his mother. It follows that all of the sperm a wasp, ant, or other hymenopteran gives to his daughters are identical. Therefore, sisters are genetically closer to one another than is the case in other kinds of animals. Three-fourths of their genes are identical instead of the usual one-half.

To see the consequences, put yourself in the place of a wasp surrounded by relatives. You are connected by one-half your genes to your mother and by the same degree to your daughters. A normal amount of solicitude toward them will be enough. But you are connected to your sisters by three-fourths of your genes. A bizarre new arrangement is now optimal: in order to insert genes identical to your own into the next generation, it is more profitable for you to raise sisters than it is to raise daughters. Your world has been turned upside down. How can you now best reproduce your genes? The answer is to become a member of a colony. Give up having daughters, and protect and feed your mother in order to produce as many sisters as possible. So the best succinct advice to give a wasp is: become an ant.

The relationship to your brothers is equally odd. They don't have the same father; in fact, they have no father at all. As a consequence they are related to you by only one-fourth of their genes. The ideal then is to raise only enough brothers and these only at times required for the insemination of young queens, and to spread some of your genes that way. An even greater indifference is optimum if you are a brother. You have the chance to father an entire new colony. It does not pay to spend time raising sisters, much less risking your life hunting for food. Better to live at the expense of the colony and specialize in both your body and behavior for the insemination of females. In short, if you are a male in a hymenopterous colony, be a drone.

Hamilton's conception seemed to explain a number of idiosyncratic facts about the societies of ants, bees, and wasps that had been staring us in the face yet had for the most part been ignored. One was the phylogenetic pattern of colonial life. Advanced social existence has arisen independently within the order Hymenoptera a dozen times, even though both the solitary and colonial forms of the order make up only 13 percent of known insect species. The only origination elsewhere was in the termites, insects descended from cockroach-like ancestors early in the Mesozoic Era. Another puzzle awaiting explanation was the role of gender in the insect societies. In hymenopteran insects males are always drones, and the workers are always female—in contrast to termites, which have an ordinary form of sex determination and, as expected, produce both male and female workers. Hamilton, in his original conception, seemed to have provided the key to many of the peculiarities of ant and other hymenopteran societies.

The story, however, does not end here. There is a twist within the twist. Robert L. Trivers, an American sociobiologist, noticed that the Hamilton argument is true only if ant workers manipulate their investment in the colony so as to expand three times more energy in the production of new queens, which are the females destined to found new colonies, than they put into the production of males. The reason is the following elementary simple arithmetical relationship (all of these important ideas could have been dashed on the back of an envelope in a few minutes): If the same number of new queens and males are produced, the overall genetic relationship between the workers and these reproductive siblings comes out to be one-half, just the same as if sex determination were by ordinary means instead of by haplodiploidy. It goes as follows: $\frac{3}{4}$ (degree of sister relationship) $\times$ $\frac{1}{2}$ (fraction of royals—queens and males—that are queens, hence sisters) $+ \frac{1}{4}$ (degree of brother relationship) $\times$ $\frac{1}{2}$ (fraction of royals that are males, hence brothers) $= \frac{1}{2}$; that is, $(\frac{3}{4} \times \frac{1}{2}) + (\frac{1}{4} \times \frac{1}{2}) = \frac{1}{2}$. The only way for the workers to promote the multiplication of their own genes is to increase the fraction of sisters, and the highest yield will come if the fraction is $\frac{3}{4}$: $(\frac{3}{4} \times \frac{3}{4}) + (\frac{1}{4} \times \frac{1}{4}) = \frac{5}{8}$. The 3:1 ratio should be in equilibrium in evolution because the expected reproductive success of the males will then be three times that of the queens on a per-gram basis.

But—can workers actually "know" that their interests are best served by investing three times as much in new queens as in new males? The data accumulated to date indicate that, somehow, they do exercise this control. And in managing it, they thwart the best interests of their mother, who would maximize her own gene duplication if the sex ratio were 1:1 instead of 3:1. The reason she should prefer 1:1 is that she is equally related to her sons and daughters, and hence distorting the ratio would cause a loss in her investment. It seems to follow that the workers run the show in ant colonies. In their readiness to sacrifice their bodies, they are still acting in the selfish interest of their genes. Darwin has the basic conception right, but he could never have foreseen the marvelous and tortuous route by which his early idea of kin selection would ultimately be upheld.

The conception is not without flaws in practical application. It works best, for example, if all members of the colony have the same father. But we know now that in a sizable minority of ant species the queen mates with two to several males, causing the workers to be less closely related. Nevertheless, it is easily possible, although still untested by experiment, that the nurse workers might express bias by raising the queens and males most closely related to them.

Other consequences follow from regarding the insect society as a product of evolution by natural selection. The concept of the selfish gene, which is seminal in the understanding of ant colonies and other close-knit animal societies, presupposes that relatives can recognize one another and discriminate against strangers. And, sure enough, it turns out that ants possess this ability to an extreme degree. They smell the difference. To see how they monitor the colony odor, watch a column of workers streaming back and forth between the nest and food. The ants meet head on and inspect one another with little or no pause, all in a split second. When the action is spread out with the aid of slow-motion cinematography, each worker can be seen to sweep her antennae over a portion of the other ant's body. In that instant the olfactory organs in the antennae tell her whether the other ant is friend or foe. If it is a friend, she runs on by without pausing. If it is a foe—a member of a different colony—she either flees the scene or halts to examine the stranger more closely. Then she may attack.

When a worker ant from one colony blunders into the nest of another, the residents immediately recognize her as a stranger. A broad spectrum of responses to such aliens is possible. At the benign end the residents accept the intruder but offer her less food until she has time to acquire the colony odor on her body. At the opposite extreme, they attack violently, locking their mandibles on her body and appendages while stinging her or spraying her with poisonous secretions.

The colony odor appears to be spread over the entire surface of each ant's body. Some evidence exists that it is a distinctive blend of hydrocarbons. These substances are the simplest of all organic compounds structurally, being composed entirely of carbon and hydrogen strung out in chains. Among the most elementary and familiar examples are methane and octane. But hydrocarbon molecules can be varied almost indefinitely by lengthening the carbon chain, by adding side chains, and by inserting double or triple bonds between the carbon atoms in place of the usual single bonds. Diversity can be expanded still further by mixing different hydrocarbons together and by shifting the proportions—in effect creating a bouquet of smells. To the human nose this blend might vaguely resemble the effluent of an automobile service station, but to the ant it exudes the subtle ambience of friendship and security. Hydrocarbons have an additional purely physical advantage: they are readily soluble in the epicuticle, the waxy film that coats the bodies of ants and other insects. As we write, the hydrocarbon hypothesis has yet to be definitively proven, but there is evidence that the substances do play at least a supporting role.

Whatever its exact chemistry, where does the colony odor originate? If every worker manufactured her own scent, the nest would be filled with a pandemonium of odors, and a tight social organization might be difficult or impossible to achieve. Colonies function efficiently to the extent that they acquire a common and distinctive blend of chemical compounds. Entomologists have suggested several ways in which ants might create the communal odor. First and most obviously, odors can be picked up from the environment, like the scent of a smoke-filled restaurant carried in the wool of a diner's coat. Members of the same ant colony regularly rub against their nestmates and lick their body surfaces. In most species, they also regurgitate liquid food stored in their chitin-lined crops. Not only can distinctive blends be created this way, but the entire nest population can share substances so extensively as to create a single colony-wide odor. That, at least, is the theory.

Another possible source of the common odor might be hereditary substances secreted from special glands in the body. Like food fragrances and other odors, these materials (if they exist) can be passed from one ant to another by grooming and regurgitation.

Whether by the acquisition of odors from the environment or by their hereditary production inside the body, the mixing of substances hour by hour ensures that the colony will possess an olfactory Gestalt, a particular common smell emanating only from that colony. The Gestalt can change as the environment or the hereditary makeup of the colony shifts. Inconstancy in the signal through time creates no great difficulty. Experiments have shown that adult ants are able to learn new colony odors, and they are especially prone to do so while they are still relatively young.

There is yet another way to create a colony odor, and it is both the simplest and most secure of all. Let the queen generate the identifying chemicals and then depend on the workers to pass them around by grooming and regurgitation. This system actually exists. It was discovered in carpenter ants of the genus *Camponotus* by Bert Hölldobler and a young co-worker, Norman Carlin. Using a series of intricate experiments, transferring queens and workers back and forth among laboratory colonies, Carlin and Hölldobler found that carpenter ants use not just the queen odor but the other two possible sources as well, and in a hierarchical manner. In particular, cues derived from the mother queen are most important to the workers in recognizing colony nestmates, followed next by substances arising from the workers, and then by odors from the environment.

Three levels of aggression recognized by Norman Carlin and Hölldobler in their study of aggression among workers of *Camponotus floridanus*. They are, from top to bottom, a simple threat display, grasping and pulling the appendages (in this case the antennae), and a full attack, which usually ends with the death of one or both of the adversaries.

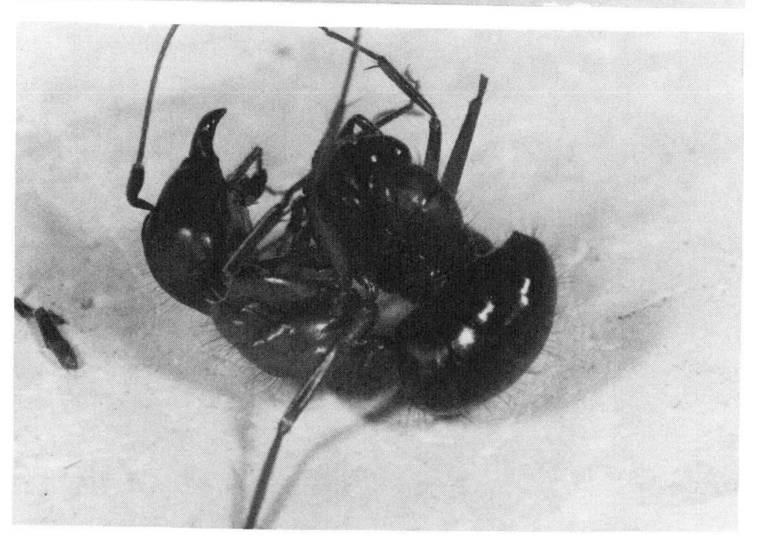

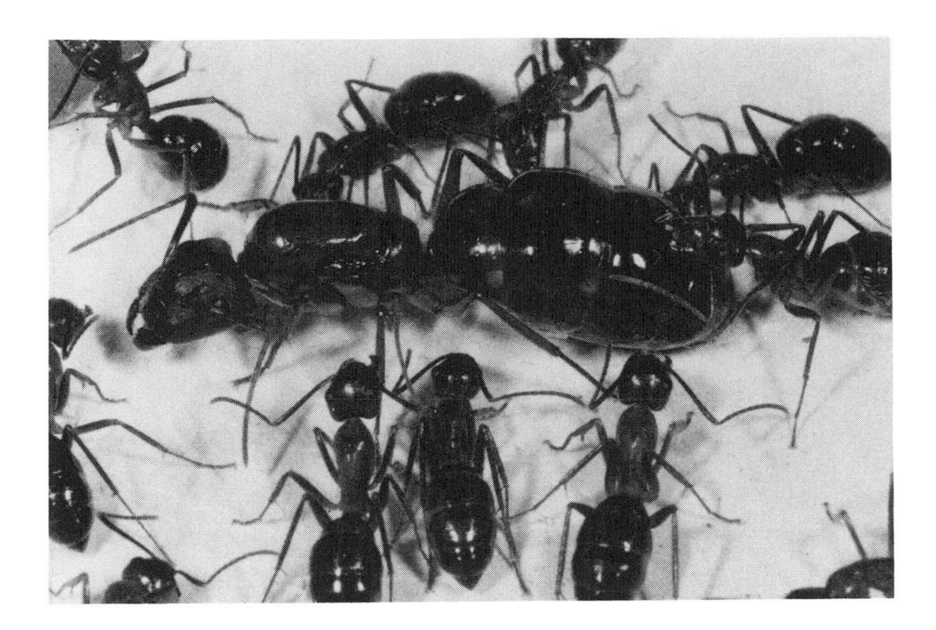

Workers of *Camponotus floridanus* surround their nest queen, licking her body almost continuously. By doing this they evidently pick up chemical queen labels that form an important component of the colony odor.

The olfactory world of the ants is as alien and complex to us as though these insects were colonists from Mars. In what may be the ultimate signature of their commitment to olfaction, they even use a small number of chemicals to recognize and dispose of corpses, while ignoring other signs of death. When an ant dies inside the nest it simply falls over, often with legs crumpled beneath it. Its nestmates at first pay no attention, because it still has approximately the right odor for a living worker. After a day or two, as decomposition sets in, other workers pick it up, carry it out of the nest, and drop it on a refuse pile. Ants, it should be noted in passing, do not have cemeteries, although some writers in ancient Greece and Rome thought they did and the myth they generated persists to this day. Corpses are merely added to the colony's garbage dump or else are dropped on bare ground away from the nest. Sometimes robber ants belonging to other species snatch the bodies away and carry them home for food.

With two fellow researchers, Wilson set out in 1958 to determine which of the chemicals of decay are used by ants to identify their dead. The collaboration was one of the first efforts to characterize the olfactory codes of these insects, and the method used was direct in the extreme. We first obtained in pure synthetic form an array of compounds known to accumulate in insect corpses; fortunately, this arcane topic of chemistry had already been carefully researched by other scientists. We daubed tiny amounts of the substances on squares of paper, which we placed inside laboratory nests of harvester ants and fire ants. We then watched to see which pieces were carried out to the refuse pile. For weeks the laboratory reeked of foul smells of the kind that emanate from dead bodies, including an unprepossessing array of fatty acids, amines, indoles, and sulfurous mercaptans. To our surprise only one small class of chemicals worked on the ants, although all worked on us, the investigators. Long-chain fatty acids alone, especially oleic acid, or their esters alone, or both together, triggered the full corpse-removal response. And when real corpses were thoroughly leached and cleansed of oleic acid with solvents, they were no longer taken out of the nest, proving that immobility alone does not a dead body make, at least not in the mind of an ant.

So far as workers are concerned, then, a corpse is defined as something with oleic acid or a closely similar substance on its body. Ants are totally narrow-minded on this subject. Their classification of a corpse extends even to living nestmates that carry the signifying odor. When we daubed a small amount of oleic acid on live workers, they were picked up and carried, unprotesting, to the refuse pile. After being dropped, they cleaned themselves and returned to the nest. If the cleaning was not thorough enough, they were carried out and dumped again.

The lessons entomologists have learned from these various studies on ants in the field and laboratory are, first, that the ability to classify other individuals quickly and precisely is crucial to social life; and, second, because this task requires the processing of a very large amount of information on odor and taste by a brain the size of a grain of salt—or even smaller—the ants must follow a set of simple, hard-and-fast rules. As a result they respond almost automatically to a predetermined set of chemicals, ignoring most of the remaining swarm of cues that human observers take for granted. Such may seem an improbable outcome of evolution, but it has worked splendidly well.

The Superorganism

All ants may look the same to the naked eye, but only for the reason that birds are hard to tell apart a mile away. Viewed close up, say 2 inches from the eye, with a hand lens to magnify them, the 9,500 or so known species of ants differ among themselves as much as do elephants, tigers, and mice. In size alone the variation is spectacular. An entire colony of the smallest ants, for example that of a *Brachymyrmex* in South America or of an *Oligomyrmex* in Asia, could live comfortably inside the head capsule of a soldier of the largest species, the giant Bornean carpenter ant *Camponotus gigas.*

Ants vary correspondingly in brain size from one species to the next, by as much as a hundredfold over all the known species. Does this mean, however, that the largest ones are more intelligent, or at least driven by a more complicated set of instincts? The answer is yes to the question of instinct (no precise measures of intelligence exist), but the difference is slight. The number of behavioral categories, comprising various acts of grooming, egg care, the laying of odor trails, and so forth, ranges from 20 to 42 across the many species in which they have been counted. The largest ants have only about 50 percent more such categories than the smallest ones. This degree of variation can be detected only through hours of meticulous recording.

In the course of evolution the brain capacity of individual ants has probably been pushed close to the limit. The amazing feats of the weaver ants and other highly evolved species comes not from complex actions of separate colony members but from the concerted actions of many nestmates working together. To watch a single ant apart from the rest of the colony is to see at most a huntress in the field or a small creature of ordinary demeanor digging a hole in the ground. One ant alone is a disappointment; it is really no ant at all.

The colony is the equivalent of the organism, the unit that must be examined in order to understand the biology of the colonial species. Consider the most organism-like of all insect societies, the great colonies of African driver ants. Viewed from afar and slightly out of focus, the raiding column of a driver-ant colony seems a single living entity. It spreads like the pseudopodium of a giant

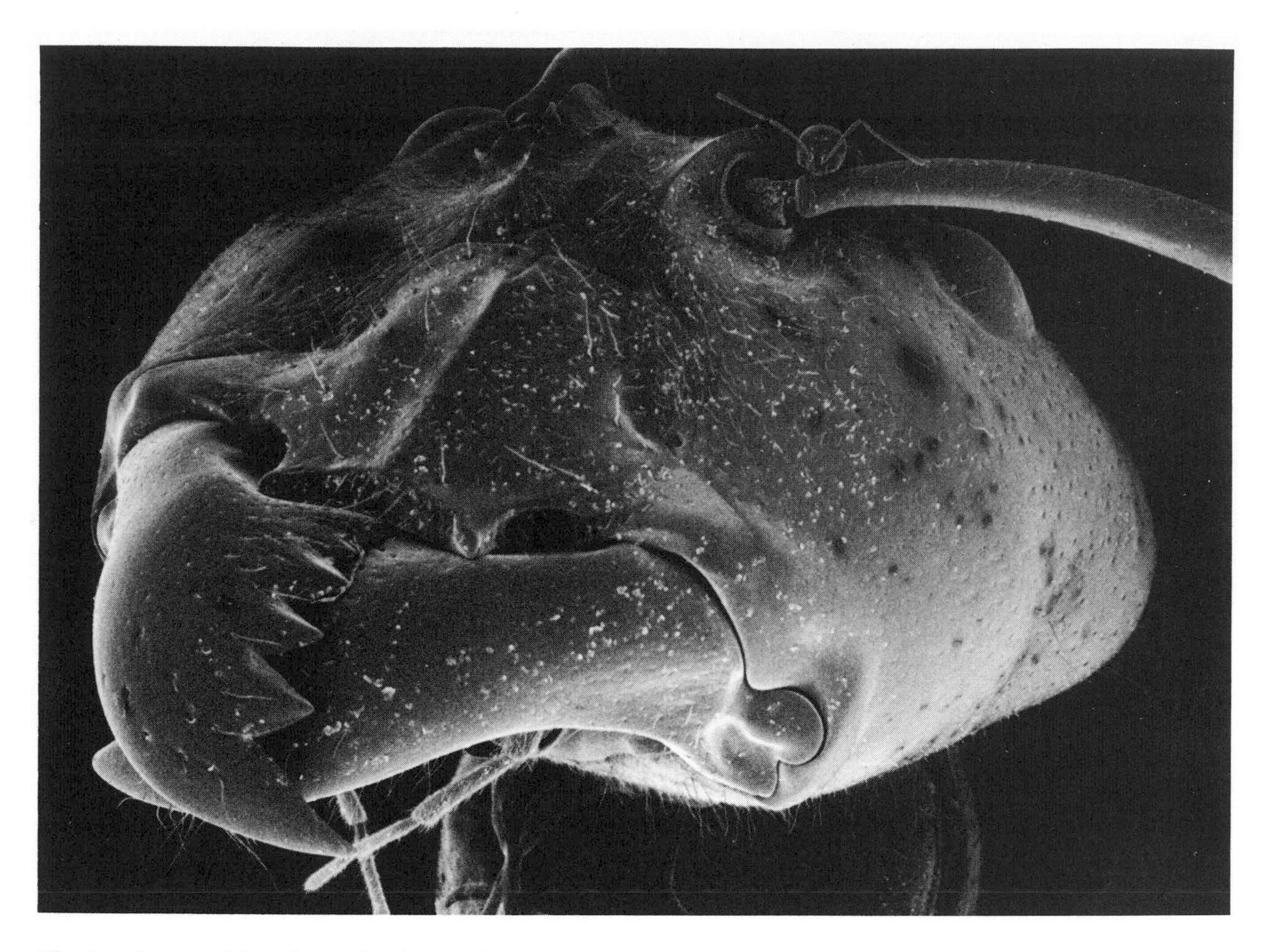

The size of ants and the colonies they form—the superorganisms—vary enormously. An entire colony of *Brachymyrmex* from South America (a worker of which is shown peeking from behind the antenna of a Bornean carpenter ant *Camponotus gigas*) would fit into the head of the larger ant. (Scanning electron micrograph by Ed Seling.)

The driver-ant superorganism: a foraging swarm of *Dorylus* driver ants in East Africa. (Drawing by Katherine Brown-Wing.)

ameba across a hundred yards of ground. A closer look reveals it to comprise a mass of several million workers running in concert from the subterranean nest, an irregular network of tunnels and chambers dug into the soil. As the column emerges, it first resembles an expanding sheet and then metamorphoses into a treelike formation, with the trunk growing from the nest, the crown an advancing front the width of a small house, and numerous anastomosing branches connecting the two. The swarm is leaderless. The workers rush back and forth near the front at an average speed of 4 centimeters per second. Those in the van press forward for a short distance and then turn back into the tumbling mass to give way to other advance runners. The feeder columns, resembling thick black ropes lying along the ground, are in reality angry rivers of ants coming and going. The frontal swarm, advancing at 20 meters an hour, engulfs all the ground and low vegetation in its path, gathering and killing almost all the insects and even snakes and other larger animals unable to crawl away. (Once in a great while the victims include a human infant left unattended.) After a few hours the direction of flow is reversed, and the column drains backward into the nest holes.

To speak of a colony of driver ants or other social insects as more than just a tight aggregation of individuals is to speak of a superorganism, and therefore to invite a detailed comparison between the society and a conventional organism. The idea—the dream—of the superorganism was extremely popular in the early part of this century. William Morton Wheeler, like many of his contemporaries, returned to it repeatedly in his writings. In his celebrated 1911 essay, "The Ant Colony as an Organism," he stated that the animal colony is really an organism and not merely the analog of one. It behaves, he said, as a unit. It possesses distinctive properties of size, behavior, and organization that are transmitted from colony to colony and from one generation to the next. The queen is the reproductive organ, the workers the supporting brain, heart, gut, and other tissues. The exchange of liquid food among the colony members is the equivalent of the circulation of blood and lymph.

Wheeler and other theorists of his day knew they were on to something important. Their voice was also within the idiom of science. Few succumbed to the mysticism of Maurice Maeterlinck's "spirit of the hive," a transcendent force that somehow emerges from, or perhaps guides, or drives, the communion of the insects. Most did not stray from the obvious physical analogies between the organism and the colony.

This exercise, however elaborate or inspirational, eventually exhausted its possibilities. The limitations of the approach based primarily on analogy became increasingly obvious as biologists discovered more of the fine details of communication and caste formation that lie at the heart of colonial organization. By 1960 the expression "superorganism" had all but vanished from the vocabulary of the scientists.

Old ideas in science, however, never really die. They only sink to mother Earth, like the mythical giant Antaeus, to gain strength and rise again. With a far greater knowledge of both organisms and colonies than was available just three decades ago, comparisons of these two levels of biological organization could be resumed with greater depth and precision. The new exercise had a goal larger than the intellectual delectations of analogy. It now aimed to mesh information from developmental biology with that from the study of animal societies to uncover general and exact principles of biological organization. The key process at the level of the organism is now seen to be morphogenesis, the steps by which cells change their shape and chemistry and move en masse to build the organism. The key process at the next level up is sociogenesis, which consists of the steps by which individuals undergo changes in caste and behavior to build the society. The question of general interest for biology is the similarities—the joint rules and algorithms—between morphogenesis and sociogenesis. To the extent that these common principles can be defined clearly, they bid fair to be recognized as the long-sought laws of general biology.

It follows that ant colonies are more than of passing interest to scientists. The ultimate possibilities of superorganism evolution are perhaps best expressed not by the driver ants, but by the equally spectacular leafcutter ants of the genus *Atta*. Fifteen species are known, all limited to the New World from Louisiana and Texas south to Argentina. With the closely related genus *Acromyrmex* (24 species, also New World), the species of *Atta* are unique among animals in their ability to grow fungi on fresh vegetation brought into their nests. They are true agriculturists.

Their crop consists of "mushrooms," which are actually masses of thread-shaped hyphae resembling bread mold. Feasting on this unlikely material, colonies reach an immense size, at maturity consisting of millions of workers. Each colony can daily consume as much vegetation as a grown cow. Several species, including the notorious *Atta cephalotes* and *Atta sexdens*, are the principal insect pests of South and Central America, destroying billions of dollars of crops yearly. But they are also among the key elements of the ecosystems. They turn over and aerate large quantities of soil in the forests and grasslands, and they circulate nutrients essential to the lives of vast assemblages of other organisms living there.

The leafcutters sustain their agriculture through a near-miraculous series of small, precise steps conducted in underground chambers. All of the species appear to follow the same basic life cycle to pass the technology across generations. It begins with the nuptial flights. Some species, such as *Atta sexdens*, hold the flights in the afternoon, while others, including *Atta texana* of the southwestern United States, conduct them in the darkness of night. With furiously beating wings, the heavy virgin queens labor upward into the air, where they meet and mate with as many as five or more males in succession. While still aloft, each queen receives 200 million or more sperm from her suitors—all of whom will die within a day or two—and stores them in her spermatheca. There they will lie inactive for as long as 14 years, the known maximum life span of queens, or even longer. One by one they will be paid out to fertilize the eggs sliding down the ovarian tubes to the outside.

In her long lifetime a leafcutter queen can produce as many as 150 million daughters, the vast majority of which are workers. As her colony grows to maturity, some of these females grow up not into workers but into queens, each capable of founding new colonies on her own. Others of her progeny arise from unfertilized eggs to become the short-lived males. All the prodigious manufacture starts when the newly inseminated queen creates the beginnings of the nest and raises her first crop of workers. She descends to the ground and rakes off her four wings at the base, rendering herself forever earthbound. She then digs a vertical shaft 12–15 millimeters in diameter straight down into the soil. At about 30 centimeters, she widens the shaft to form a room 6 centimeters

Newly inseminated queens of the leafcutter-ant genus *Atta* start new nests by digging a vertical shaft in the soil *(A)*. They raise the first fungal garden by applying drops of anal fluid to the clump of hyphae *(B)*. Three stages in the subsequent growth of the garden and worker brood are depicted in *C*. (Drawing by Turid Forsyth.)

across. Finally, she settles into the chamber to cultivate a new garden of fungi and rear her brood.

But wait—how can the queen raise a garden if she left the symbiotic fungus behind in the mother nest? No problem—she did not leave it behind. Just before the nuptial flight she tucked a wad of the threadlike hyphae into a small pocket in the bottom of her mouth cavity. Now she spits out the packet onto the chamber floor. Her garden started, she soon afterward also lays 3 to 6 eggs.

At first the eggs and the little fungus garden are kept apart, but by the end of the second week, when more than 20 eggs have accumulated and the fungal mass is ten times the original size, the queen brings the two together. At the end of the first month the brood, now consisting of eggs, larvae, and the first of the pupae, is embedded in the center of a mat of proliferating fungi. The first adult workers emerge 40 to 60 days after the first eggs were laid. During all this time the queen cultivates the fungus garden herself. At intervals of an hour or two she tears out a small fragment of the garden, bends her abdomen forward between her legs, touches the fragment to the tip of the abdomen, and impregnates it with a clear yellowish or brownish droplet of fecal liquid. Then she returns the fragment to the garden. Although the queen does not sacrifice her own eggs as a culture medium for the fungus, she does consume 90 percent of the eggs herself. And when the larvae first hatch, they are fed with eggs thrust directly into their mouths.

During all this time the leafcutter queen subsists entirely on energy obtained from the breakdown and metabolism of the wing muscles and fat within her own body. She grows lighter by the day, caught in a race between starvation and the creation of a force of workers adequate to prolong her life. When the first workers do appear, they begin to feed on the fungus. After about a week they dig their way up through the clogged entrance channel and start foraging on the ground in the immediate vicinity of the nest. They bring in bits of leaves, chew them into pulp, and knead them into the fungus garden. About this time the queen ceases attending both brood and garden. She turns into a virtual egg-laying machine, a condition in which she is destined to remain the remainder of her life.

The colony is now self-supporting, with an economy based on the harvesting of outside materials. At first it expands only slowly. Then during the second and third years its growth accelerates quickly. Finally, it tapers off as the colony starts to produce winged queens and males, which are released during the nuptial flights and hence contribute nothing to the communal labor.

The ultimate size of mature leafcutter colonies is enormous. The record may be attained by *Atta sexdens* at 5 to 8 million. One nest excavated in Brazil comprised over a thousand chambers varying in size from a closed fist to a soccer ball, of which 390 were filled with fungus gardens and ants. The loose soil that had been brought out and piled on the ground by the ants, when shoveled off and measured, occupied 22.7 cubic meters (800 cubic feet) and weighed approximately 40,000 kilograms (44 tons). The construction of one such nest is easily the equivalent, in human terms, of building the Great Wall of China. It requires roughly a billion ant loads to build, each weighing four or five times as much as a worker. Each load was hauled straight up from depths in the soil equivalent, again in human terms, to as much as a kilometer.

The routines of the leafcutters are among the great wildlife spectacles of the New World tropics. Every field biologist is drawn to grandeur on this scale, even though the actors are minute in size. During his first trip to the Brazilian Amazon, in the rain forest near Manaus, Wilson was spellbound by the sight of one of the foraging expeditions of *Atta cephalotes.* At dusk on the first day in camp, as the light failed to the point where he and his companions found it difficult to distinguish small objects on the ground, the first worker ants came scurrying purposefully out of the surrounding forest. They were brick red in color, about 6 millimeters in length, and bristling with short, sharp spines. Within minutes several hundred had entered the campsite clearing and formed two irregular files that passed on either side of the biologists' shelter. They ran in nearly straight lines across the clearing, their paired antennae scanning right and left, as though drawn by some directional beam on the far side of the clearing. Within an hour, the trickle expanded to twin rivers of tens of thousands of ants running ten or more abreast. The columns could be traced back to their source easily with a

The architecture of a mature nest of the leafcutter ant *Atta vollenweideri* in Paraguay. Garden chambers contain the growing masses of fungus, on which the ants feed, and dump chambers are filled with exhausted vegetation substrate on which the fungus has been growing. (Modified by N. A. Weber from an illustration by J. C. M. Jonkman in L. A. Batra, ed., *Insect-Fungus Symbiosis: Mutualism and Commensalism,* Montclair, N.J., Allanheld and Osman, 1979.)

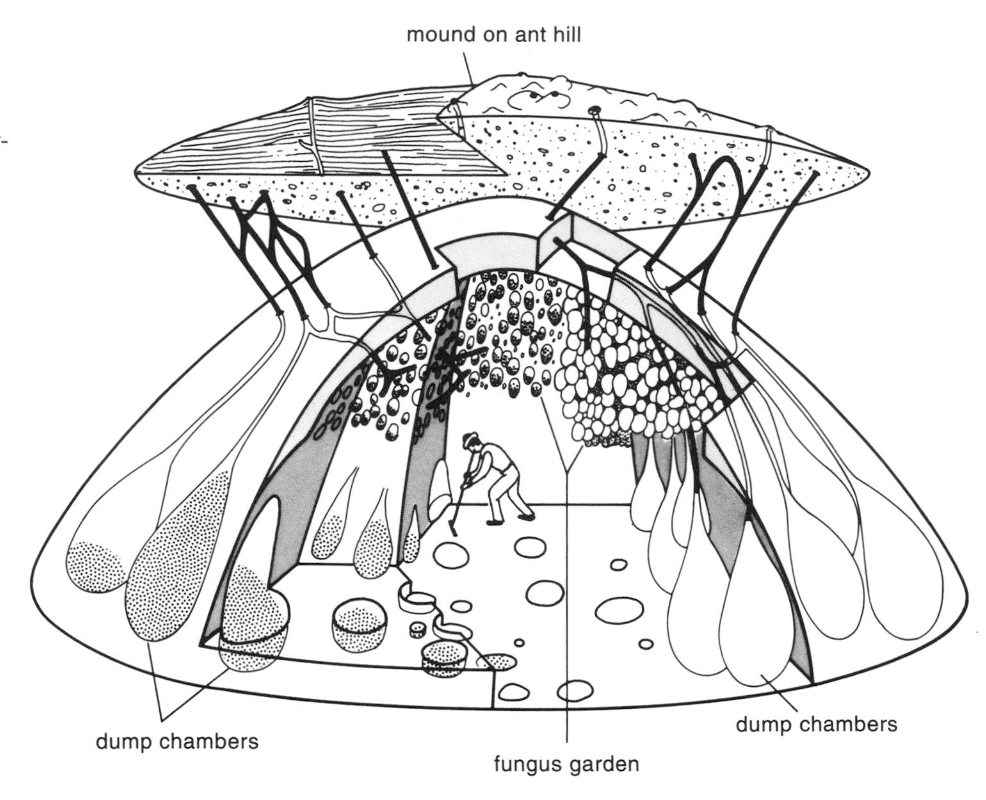

flashlight. They came from a huge earthen nest a hundred meters from the camp up an ascending slope, crossed the clearing, and disappeared again into the forest. By climbing through tangled undergrowth Wilson and his companions were able to locate one of the main targets, a tall tree bearing white flowers high in its crown. The ants streamed up the trunk, scissored out pieces of leaves and petals with their sharp-toothed mandibles, and headed home carrying the fragments over their heads like little parasols. Some of the workers dropped their pieces to the ground, apparently deliberately, where they were picked up and carried away by newly arriving nestmates. At maximum activity, shortly after midnight, the trails were a tumult of ants bobbing and weaving past one another like miniature mechanical toys.

Ant workers usually do not exhibit any dominance or conflict behavior; they cooperate in the care of the offspring of their mother queen, as shown in this portrait of the Neotropical ponerine species *Ectatomma ruidum*. Nevertheless, conflict among nestmate workers can arise if the queen is lost and the workers become fertile.

A dominance display between two queens of *Myrmecocystus navajo* is shown in the upper picture. The dominant individual steps onto the back of the subordinate, who shows her submission by crouching and opening her mandibles. In most cases only one queen succeeds in becoming the mother of a mature colony. The lower picture shows a successful foundress of *Myrmecocystus mexicanus* with her first offspring, including larvae, pupae, and young adult workers.

The nest chamber of a colony of the ponerine species *Harpegnathos saltator.* The picture was taken near Jog Falls, India, where colonies of this unusual ant are quite common.

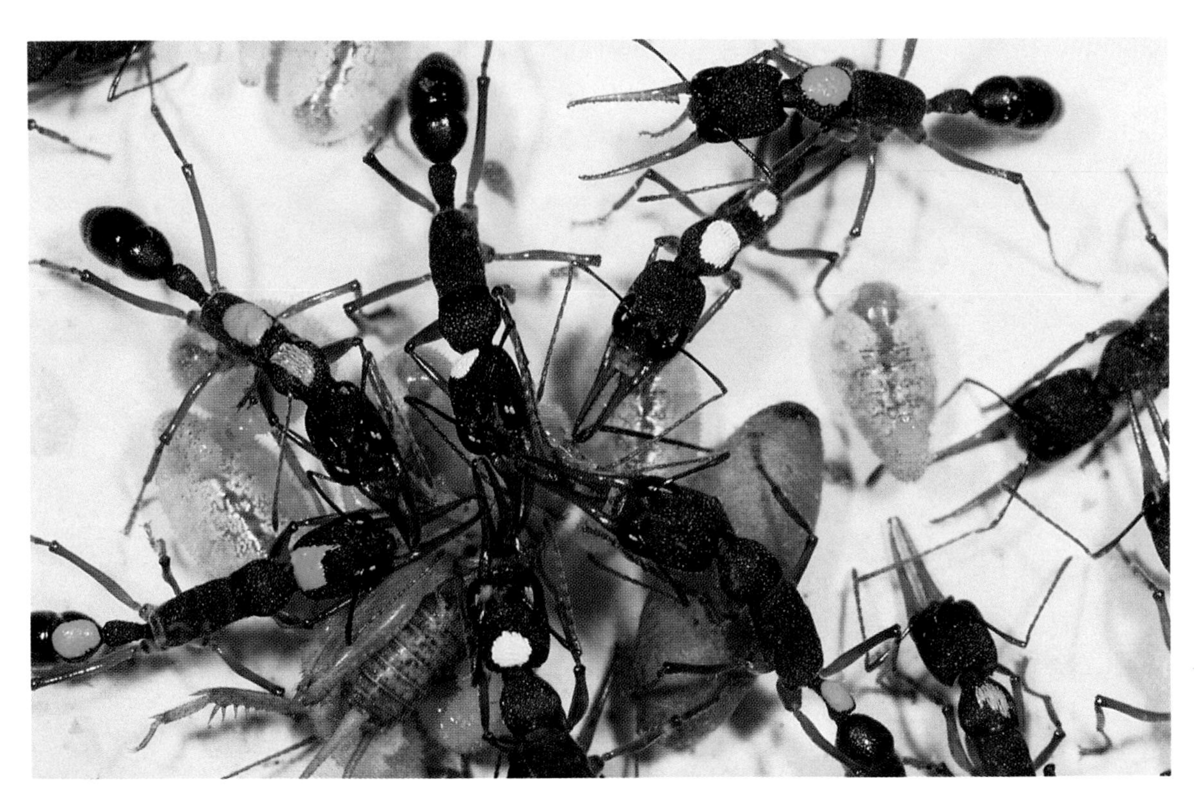

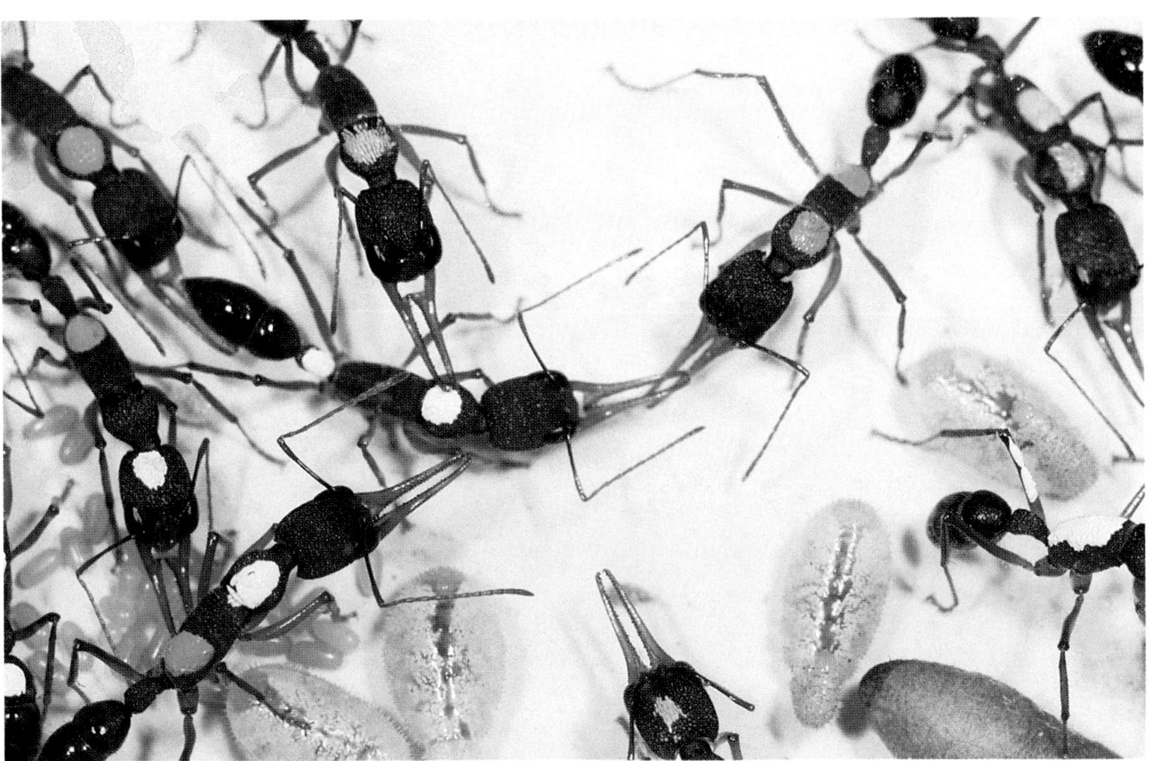

Facing page
For close investigations of individual behavior, whole colonies of *Harpegnathos saltator* were brought into the laboratory, and each ant was then marked with an individual color-code. The upper picture shows several individuals sharing a prey. The lower shows two ants (in the center) engaged in a dominance duel, during which they confront each other head on and alternately lash each other with their antennae.

Workers of neighboring colonies of the harvester ant *Pogonomyrmex barbatus* often engage in fierce combat when defending their territories *(upper right).* The ants commonly fight to the death, as shown in the photograph of a *P. barbatus* forager carrying the head of a former opponent attached to her waist by its viselike mandibles *(lower right).*

A truly universal feature in the social life of ants is brood care. Here workers of *Camponotus planatus* groom, feed, and protect the larvae and pupae of the colony.

Facing page
Mutual grooming *(above)* and social food sharing by regurgitation *(below)* are altruistic acts nearly universal in ant societies. They are displayed here by workers of the predatory South American ant *Daceton armigerum.*

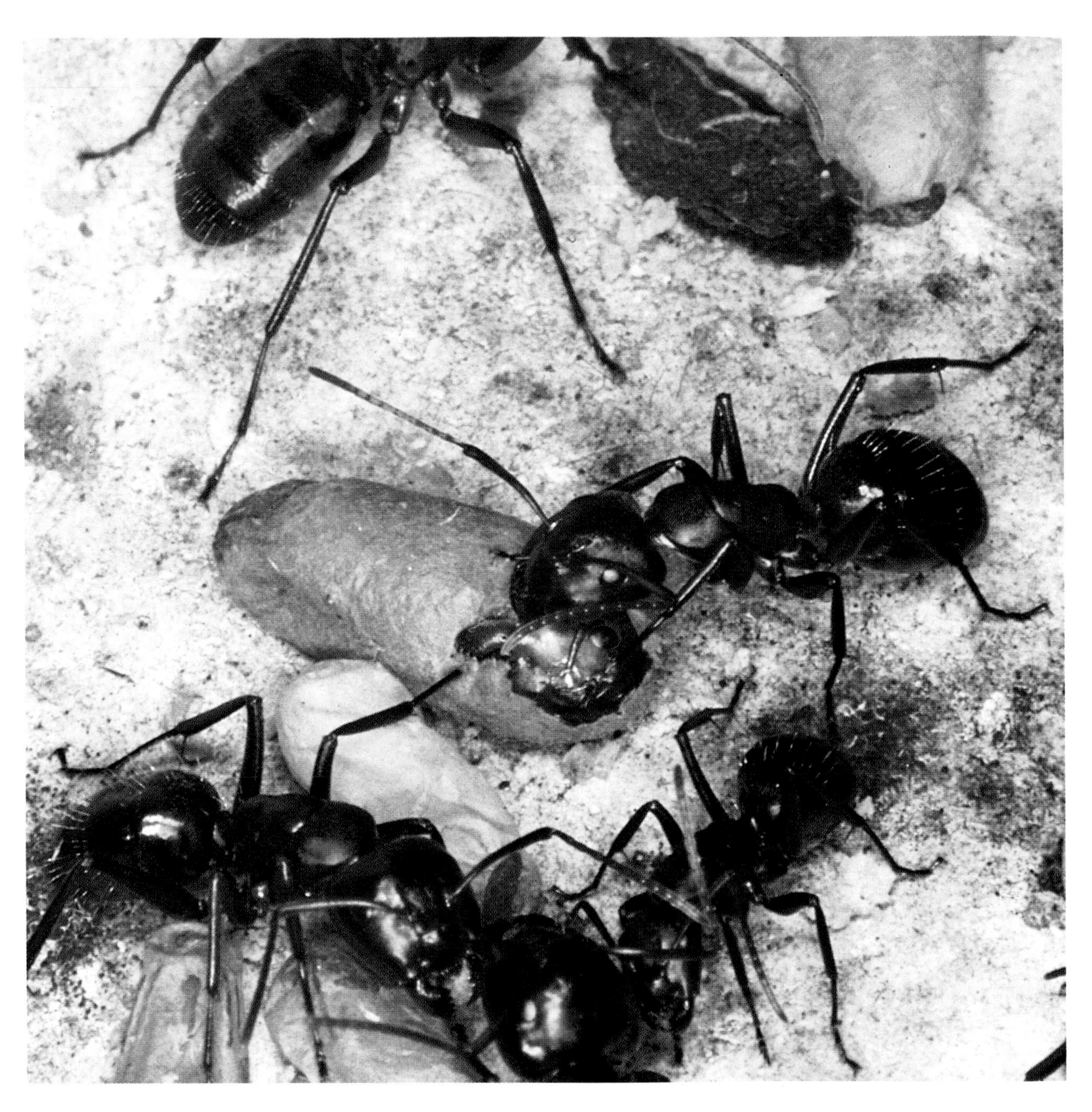

In many ant species newly emerging adults are unable to free themselves from their pupal cocoon. In this episode a *Camponotus ligniperda* worker helps a young nestmate to free itself.

For many visitors to the forest, even experienced naturalists, the foraging expeditions are the whole of the matter, and individual leafcutter ants seem to be inconsequential ruddy specks on a pointless mission. But a closer look transforms them into beings of another order. If we magnify the operation to human scale, so that an ant's 6-millimeter length grows into a meter and a half, the forager runs along the trail for a distance of about 15 kilometers at a velocity of 26 kilometers an hour. Each successive mile (to convert to familiar Anglo-American sports distances) is covered in 3 minutes and 45 seconds, about the current human world record. The forager picks up a burden of 300 kilograms or more and speeds back to the nest at 24 kilometers an hour—hence 4-minute miles. This fast marathon is repeated many times during the night and in many localities on through the day as well.

To follow the process to completion, and analyze the *Atta* superorganism in greater detail, Wilson set up colonies in the laboratory inside plastic chambers aligned in interconnected rows, allowing him to look deep inside the fungus gardens. He discovered that gardening is achieved by means of an intricate assembly line, in which the leaves and petals are processed and the fungus reared in steps.

Each of the steps is accomplished by a different caste. At the end of the trail, the burdened foragers drop the leaf sections onto the floor of a chamber, to be picked up by workers of a slightly smaller size who clip them into fragments about a millimeter across. Within minutes still smaller ants take over, crush and mold the fragments into moist pellets, and carefully insert them into a pile of similar material. This mass, the local garden, is riddled with channels and looks something like a gray bath sponge. Fluffy and delicate, it is easily torn apart in the hands. On the surface of its tortuous channels and ridges grows the symbiotic fungus which, along with the leaf sap, forms the ants' sole nourishment. The fungus spreads across the kneaded vegetable paste like bread mold, sinking its hyphae into the material to digest the abundant cellulose and proteins held there in partial solution.

The gardening cycle proceeds. Worker ants even smaller than those just described pluck strands from places of less dense growth and place them on the newly constructed vegetable-paste substrate. Finally, the very smallest and most abundant workers patrol the beds of fungal strands, delicately probing them with their antennae, licking their surfaces clean, and plucking out the spores and hyphae of alien species of mold. These laboring dwarfs are able to travel through the narrowest channels deep within the garden masses. From time to time they pull tufts of fungus loose and carry them out to feed their larger nestmates.

The leafcutter economy is organized around this division of labor based on size. The foraging workers, about as big as houseflies, can slice leaves, but are too bulky to cultivate the near-microscopic fungal strands. The tiny gardener workers, somewhat smaller than the printed capital letter I on this page, can grow the fungus but are too weak to cut the leaves. So the ants form an assembly line, each successive step being fashioned by correspondingly smaller workers, from the collection of pieces of leaves out of doors to the manufacture of leaf paste to the cultivation of dietary fungi deep within the nest.

The defense of the colony is also organized according to size. Among the scurrying workers can be seen a few soldier ants, 300 times heavier than the gardener workers, and with heads 6 millimeters across. Like the *Pheidole* soldiers we described earlier, these giants use sharp mandibles to clip enemy insects into pieces. They can cut through leather and slice open human skin with equal facility. When entomologists digging into a nest take no precautions, their hands are nicked all over as though pulled through a thorn bush. We have occasionally had to pause to staunch the flow of blood from a single bite, impressed by the fact that a creature one-millionth our size could stop us with nothing but its jaws.

The leafcutter colony expands to its mighty force, from giant soldiers to swarming Lilliputian gardeners, through an exactly controlled trajectory of life stages. In the first crop of adult workers reared by the queen there are no soldiers or larger-sized foraging workers. Only the smallest foragers, plus the still smaller workers needed to process vegetation and raise fungi, are present. As the colony prospers and its population grows, the size range of the worker expands to include larger and larger forms. Finally, when the population reaches about a hundred thousand, the first full-sized soldiers are added.

Wilson saw in the regularity in the growth of leafcutter colonies a

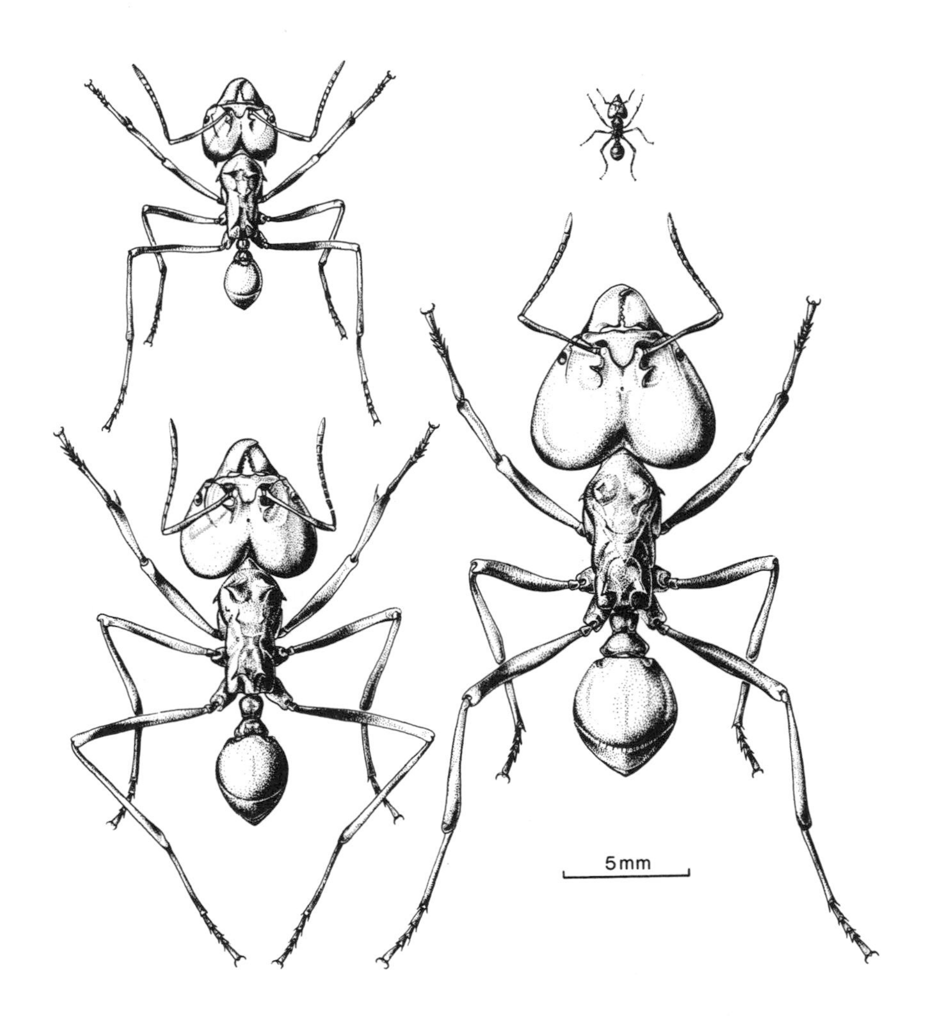

The caste system of *Atta* leafcutters, among the most complex in the social insects. The workers depicted here, from tiny gardeners to giant soldiers, are all from a single colony of *Atta laevigata*. (Drawing by Turid Forsyth.)

means to test the superorganism concept. His attention was drawn especially to the plight of the founding queen. This large ant supports herself and raises her first brood of workers by converting her body fat and wing muscles into energy. With her resources rapidly running down over a period of weeks, she must create a perfectly balanced work force on her first try. There is no room for error. In order for the first crop of workers to take over the whole agricultural task and bring food to her exhausted body, they must include in their ranks a number of tiny fungus gardeners, plus some individuals of each of the intermediate sizes required to build the leaf-paste garden, plus a few workers large enough to forage away from the nest and cut leaves.

If the queen fails to rear any of these critical sizes among her workers, the little colony dies. If she raises a soldier, or even a larger-sized foraging worker, so much of her resources will be consumed that not all the smaller castes will be affordable, and the colony will die. Wilson found that the smallest successful foragers (those able to cut through leaves of ordinary thickness) have head widths of 1.6 millimeters; in larger colonies, many of the foragers have heads twice that large, and therefore are several times heavier (and more expensive to make) than is absolutely necessary. The gardeners have heads of minimum size: 0.8 millimeters wide.

So it is clear what the founding queen must do: raise workers in her first brood whose heads vary from 0.8 to 1.6 millimeters, with a more or less even sprinkling of sizes in between. She must be careful not to omit any of these size categories, and not to go over 1.6 millimeters. And that is exactly what she does. Incipient colonies, whether they are dug up in the field for examination or cultured in the laboratory, always (at least in the cases Wilson studied) raise a crop of workers with head widths evenly distributed from 0.8 to 1.6 millimeters. Only an occasional queen creates a 1.8-millimeter worker, a risk to survival but not fatal. Larger workers never appeared in the study sample.

What is the nature of this superorganismic control? Does it come from the age of the queen and colony—or from the population size of the colony? In order to find out, Wilson let four colonies of leafcutters grow in the laboratory three to four years, at which time the worker populations reached approximately 10,000. Large foragers and even a few smaller soldiers had appeared. Next he trimmed the colonies back to a little over 200 workers, adjusting the size cohorts so that the relative numbers of workers in each were the same as in a very young colony. So the queen and colony members were now chronologically old, but the superorganism—in its size and caste configuration—was young. It had been "reborn." What would be the configuration produced by the next crop of workers? Would the sizes of the workers be those of a small colony, or would they continue on like those of the large colony before it was trimmed?

Answer: the configuration followed was that of a small colony. In other words, the size of the colony, not its age, determines the caste distribution. The experimental colonies, in one sense truly reborn, started out anew on their tightly controlled path of growth and differentiation. Had they not done so, they might not have survived. The feedback mechanism behind this remarkable control remains to be explored.

The rejuvenation of the leafcutter colony, together with other experiments on different species by other investigators, have rendered the superorganism concept more robust. They have given validity to the idea of the ant colony as a tightly regulated unit, a whole that indeed transcends the parts. And in the reverse direction, the superorganism has stimulated new forms of research. In the study of biological organization, the ant colony offers certain advantages over ordinary organisms. Unlike an organism, it can be torn apart into smaller groups that differ by age or size. These fragments can be studied in isolation, then reassembled into the original whole, with no harm done. The next day the same colony can be vivisected in yet another way, then restored to the original state—and so on. The procedure has enormous advantages. It is first of all quick and technically easy compared with analogous experiments on organisms. But it also provides its own elegant experimental control: by using the same colony repeatedly, researchers eliminate variations due to genetic differences or prior experience.

The advantage of tearing apart the colony and reassembling it repeatedly is the same as, say, vivisecting a human hand and restoring it repeatedly without pain or inconvenience, in order to discover the ideal anatomical conformation. Put more precisely, the procedure is used to learn whether the five-fingered hand humans possess is the best arrangement possible. One day we cut off the thumb (painlessly), ask the subject to perform manual tasks such as writing or opening bottles, and at the end of the day stick the thumb back on to resume its former function. The next day the terminal digits are trimmed off, and the next an extra finger is added, and so on through large numbers of arrangements.

Wilson looked at the castes of leafcutter ants as though they were fingers on a hand. He noticed that the most common group of workers that foraged away from the nest to harvest leaves and flowers have heads 2.0 to 2.4 millimeters wide. Is this the best caste for the job, the one that gathers the most vegetation with the least expenditure of energy? Wilson tested this hypothesis, and with it the implicit assumption that the caste system evolved by natural selection, by vivisecting the colony in the following manner. Each day foragers and their attendants left the laboratory nest to travel into a walled-in open space provisioned with fresh leaves. As the column of eager workers pressed through the exit, he removed all but a particular size class, such as those with head widths of 1.2, or 1.4, or 2.8 millimeters, or any other size chosen randomly on that occasion. The colony was thus transformed into a pseudomutant, a simulated mutation of the superorganism, identical in all respects to the "normal" colony (itself on other days, when the foragers were not modified) except that it was sending out a restricted, often very peculiar, stream of foragers. The leaves harvested by each pseudomutant variant were weighed, and the oxygen consumed by the ants during harvesting was measured. By these criteria the most efficient group proved to be the workers whose heads were 2.0 to 2.2 millimeters in width, the size class actually committed to the task of foraging by the colony. The leafcutter colonies, in short, do precisely the right thing for their own survival. Guided by instinct, the superorganism responds adaptively to the environment.

Social Parasites: Breaking the Code

THE GREAT STRENGTH of ants is their ability to create tight bonds and complex social arrangements with tiny brains. They have done so by cueing their behavior to a limited array of very specific stimuli. A certain terpene forms an odor trail, a tap on the lower mouthparts asks for food, a fatty acid identifies a corpse, and so on through a few dozen such signals to lead the single ant through her daily social rounds.

The superstructure of organization ant colonies have erected is impressive, but the basis of the strength—the concatenation of simple cues—is also a source of major weakness. Ants are easily fooled. Other organisms can break their code and exploit the social bond merely by duplicating one or several key signals. The social parasites that accomplish this feat are like burglars who enter a house quietly by punching the correct four or five numbers to turn off the alarm system.

Human beings are extremely hard to deceive in face-to-face encounters. They recognize a friend or member of the family by a vast number of nuanced cues, including the exact conformation of height, posture, facial features, inflection of voice, and casual mention of a mutual acquaintance. An ant recognizes a family member—a nestmate—solely by its odor, which may comprise no more than a blend of a few hydrocarbons on the body surface. Many socially parasitic beetles and other insects, a majority of which are radically different in shape and size, have mastered the art of acquiring the colony odor or the attractive scent of an ant larva. Despite the fact that they can pass no other conceivable test of recognition, they are readily admitted to the company of the ants, who are then inclined to feed, wash, and carry them bodily from place to place. To paraphrase William Morton Wheeler, it is as though a human family were to invite gigantic lobsters, midget tortoises, and similar monsters to dinner, and never notice the difference.

Some of the most sophisticated social parasites are ants that victimize other kinds of ants. The ultimate example may be *Teleutomyrmex schneideri,* a rare species discovered by the distinguished Swiss myrmecologist Heinrich Kutter. This extraordinary parasite lives ex-

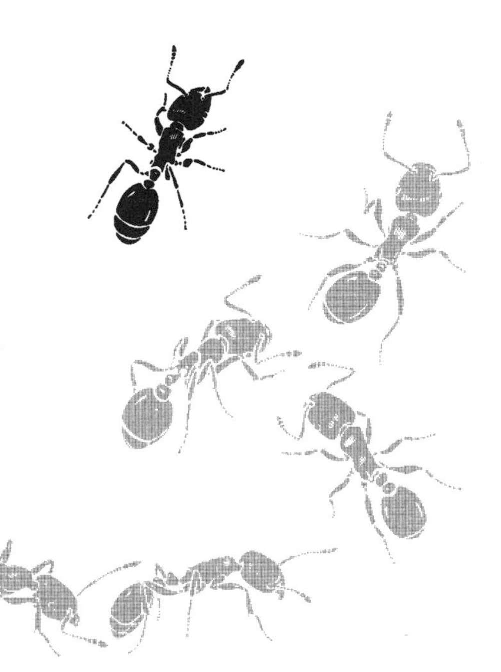

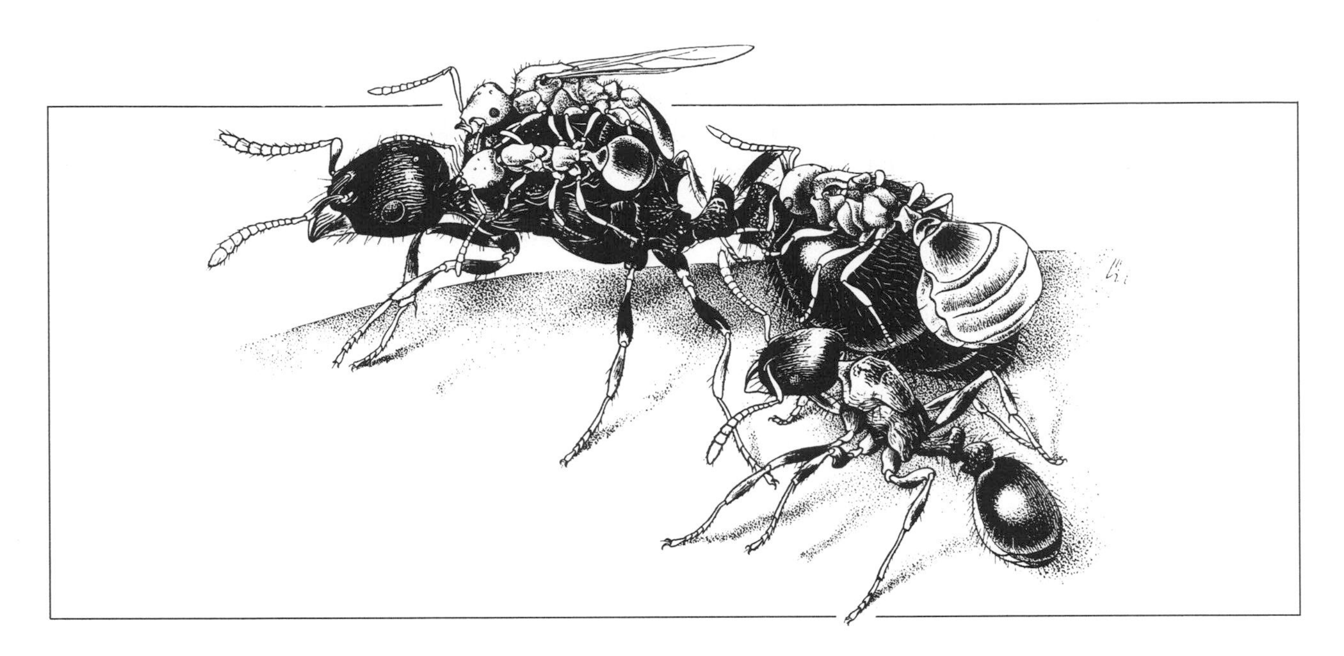

The extreme social parasite *Teleutomyrmex schneideri* in residence with its host *Tetramorium caespitum.* On the left, the two *Teleutomyrmex* queens sitting on the thorax of a host queen have not yet undergone ovarian development, and their abdomens are consequently flat and unexpanded. One still bears her wings and is almost certainly a virgin. The third *Teleutomyrmex* queen, riding on the abdomen of a host queen, has an abdomen swollen with hyperdeveloped ovarioles. A host worker stands in the foreground. (Drawing by Walter Linsenmaier.)

clusively as the guest of another ant species, *Tetramorium caespitum,* in the French and Swiss Alps. Appropriately, the Greek base of the word *Teleutomyrmex* means "final ant." The species lacks a worker caste, depending on the host workers for their care. The queens, which are tiny compared with most ants, averaging only 2.5 millimeters in length, contribute in no productive way to the economy of the host colonies. They are unique among all known social insects in being not just parasitic but ectoparasitic—they spend much of their time riding on the backs of their hosts. This peculiar habit is made possible not only by the small size of the *Teleutomyrmex* but also by their body shape. The lower surface of the abdomen (the large terminal part of the body) is strongly concave, making it possible for the parasites to press their bodies close to those of the hosts. The pads and claws of their feet are proportionately large, permitting the *Teleutomyrmex* to secure a strong purchase on the smooth chitinous body surface of other ants. The queens have a strong instinctive tendency to grasp objects, preferably the mother queen of the host colony. As many as eight have been observed riding on a single host queen, their crowded bodies and clutching legs cloaking her body and preventing her from moving.

These ultimate parasites have infiltrated the *Tetramorium* societies in every detail. They are fed through direct regurgitation from the workers. They are also permitted to share in the liquid passed to the host queen. Thus succored like cherished infants, the *Teleutomyrmex* queens are prodigiously fertile. Older individuals, their gasters swollen by masses of ovaries, lay an average of two eggs every minute.

The population of host workers is diminished by the burden of the parasite population. They nevertheless care for the *Teleutomyrmex* in every detail and rear a large crop capable of infesting other colonies nearby. At every stage of the life cycle, from egg to adult, the *Teleutomyrmex* send signals, mostly chemical in nature, that cause their hosts to accept them as full colony members.

A price has been paid during the course of evolution for this achievement in perversity. The mark of the parasite is upon the *Teleutomyrmex;* their bodies are weak and degenerate. They lack some of the glands that other ants use to produce food for larvae and protection against bacteria. Their exoskeleton is thin and lightly pigmented, their sting and poison glands are reduced in size, and their mandibles are too small and weak to handle anything but liquid food. The brain and central nerve cord are small and simplified; there is no evidence that the adults can do more than mate, fly short distances, cling to their hosts, and beg. When separated from their hosts, they do not live for more than a few days.

Teleutomyrmex schneideri, the European symbiotic wonder, is also one of the rarest ants in the world. Other extreme parasites among the ants, those completely dependent on their hosts for every detail of their care, are similarly very rare. There are no exceptions. In fact, the discovery of such parasites in a host colony, whether a new species or one already known from past collections, is a notable event for myrmecologists. They write a short article about the discovery or at least pass along the news as trade gossip. The unquestioned champion discoverer of the social parasites is Alfred Buschinger of Germany. With his teams of students and collaborators, he has hunted them around the world, illuminating the innermost secrets of their clandestine existence.

As Buschinger and others have concluded, it cannot be proved that parasitic species are short-lived, destined to descend to extinction soon after falling upon the charity of another species. But they certainly become scarce and also often restricted enough in geographic range to flirt with extinction. As in the affairs of human beings, knaves must always be less numerous than fools, else they are deprived of their livelihood.

Another well-known form of parasitism in ants is the enslavement of other kinds of ants. The dependency on forced labor is considerable, but there is far less degeneracy in anatomy and behavior. Earlier we described how colonies of honeypot ants often overrun weaker colonies, destroying the queens and capturing the younger workers and members of the honeypot castes, who then live and work in the conquerors' nests. This is true slavery even by the strictest definition: subjugation and forced labor of members of the same species. A much more common occurrence is the enslavement by ants of members of different species. Here the terms slavery and slave-making are employed loosely. The activity is more akin to the capture and domestication of dogs and cattle by humans. Yet the term "slavery" is so well entrenched, and the behavior of the slave-makers so striking and familiar to entomologists, that we will continue to use it here. Even specialists on ant behavior prefer it to the technical expression dulosis, which has been introduced to encompass enslavement across species and pops up occasionally in entomological journals.

Nothing in the ant world is visually more striking than the slave raids of species in the genus *Polyergus,* the so-called amazon ants. Shiny red or jet black in color, large in size, bold and dashing in war, amazon ants are at the pinnacle of the slave-holding way of life. The raids are directed against the abundant colonies of similar-appearing ants in the genus *Formica.* The European species *Polyergus rufescens* is quite common in

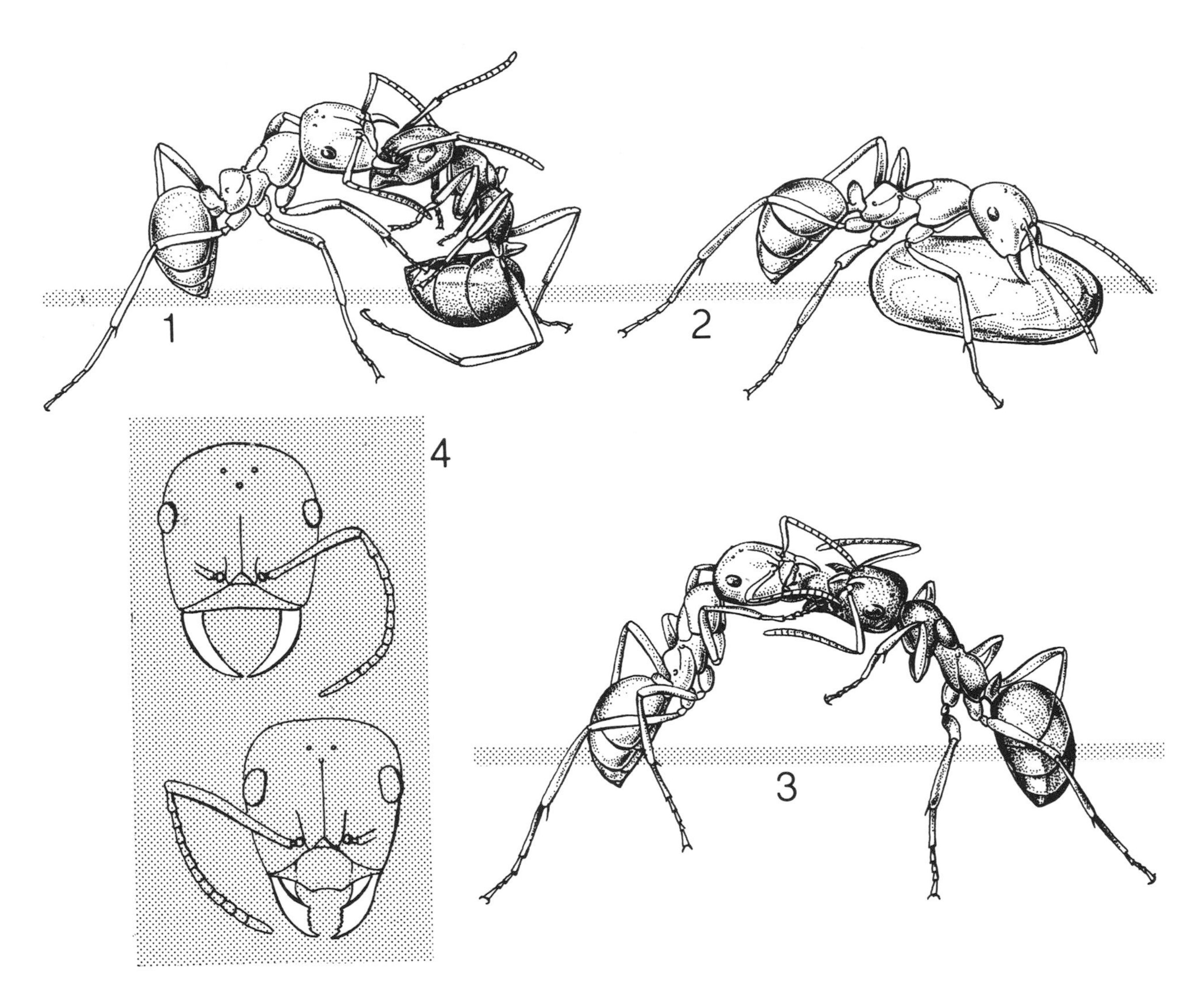

Scenes from the daily life of the European amazon ant *Polyergus rufescens.* *(1)* A *Polyergus* worker attacks a defending worker of *Formica fusca* during a slave raid, then *(2)* returns to her own nest with a *Formica* pupa (encased in a cocoon). *(3)* A *Polyergus* worker is fed by a *Formica* slave that has emerged from a captured pupa. In *(4)* the sickle-shaped mandibles of *Polyergus (above)* are contrasted with the ordinary broad mandibles of *Formica subintegra,* which uses chemical sprays rather than piercing bites to subdue defenders (see the illustration on page 131).
(Drawing by Turid Forsyth.)

the limestone habitats along the Main River near Würzburg. As a high school student, about 15 years in age, Bert Hölldobler observed many raids of the amazon ants and took detailed notes on the behavior of the robbers and their slave ants. He later learned that most of his discoveries had already been reported by the Swiss entomologist Pierre Huber in 1810 and by Auguste Forel, the great Swiss neuroanatomist, psychiatrist, and myrmecologist, in his major monograph *Le monde social des fourmis.*

Polyergus are true parasites. Fighting is the only thing they do well, as described by William Morton Wheeler: "The worker is extremely pugnacious, and, like the female, may be readily distinguished by its sickle-shaped, toothless, but very minutely denticulate mandibles. Such mandibles are not adapted for digging in the earth or for handling thin-skinned larvae or pupae and moving them about in the narrow chambers of the nest, but are admirably fitted for piercing the armor of adult ants. We find therefore that the amazons never excavate nests nor care for their own young. They are even incapable of obtaining their own food, although they may lap up water or liquid food when this happens to come in contact with their short tongues. For the essentials of food, lodging and education they are wholly dependent on the slaves hatched from the worker cocoons that they have pillaged from alien colonies. Apart from these slaves they are quite unable to live, and hence are always found in mixed colonies inhabiting nests whose architecture throughout is that of the slave species. Thus the amazons display two contrasting sets of instincts. While in the home nest they sit about in stolid idleness or pass the long hours begging the slaves for food or cleaning themselves and burnishing their ruddy armor, but when outside the nest they display a dazzling courage and capacity for concerted action" (*Ants: Their Structure, Development, and Behavior,* New York, Columbia University Press, 1910).

This concerted action, a raid by amazon ants, is a spectacular drama. Workers pour out of the nest to form a compact column running over the ground at 3 centimeters a second—the equivalent of a human brigade traveling at 26 kilometers (16 miles) an hour. When they reach their target, a nest of *Formica* ants, often 10 meters or more away, they charge into the entrance without hesitation, seize the cocoon-covered pupae, speed out again, and return to their own nest. They attack and kill any worker that opposes them, piercing the heads and bodies of the defenders with their saber-shaped mandibles. Once home, they turn the pupae over to the adult slaves for further care, and revert to their usual indolence.

The means by which the *Polyergus* workers are able to run straight to the victim colony was for many years one of the classic problems of myrmecology. While watching amazon nests in Michigan in 1966, Mary Talbot noticed that before the onset of each raid, several scout workers explored the terrain surrounding the vicinity of the particular *Formica* nest that was later attacked. The beginning of each raid was signaled by the appearance of a scout returning from the direction of the target nest. Because amazon raids did not appear to be guided by leaders, it seemed logical to Talbot that the scout must direct her nestmate by laying an odor trail from the target all the way home. It was as though the scout said, "I'm telling you there is a nest out there. Just follow the trail." How can such a hypothesis be tested? Talbot decided to speak to the amazon raiders directly, to give them instructions of her own. Using an artist's brush, she laid down dichloromethane extracts of whole *Polyergus* bodies in artificial trails leading away from the amazon nests, at the time of day raids normally occur. The attempt succeeded to a startling degree. *Polyergus* workers obediently poured from the nest and followed the trails to the end. Thus Talbot was able to activate the raid swarms at will and lead them to targets of her choosing. Finally, she induced a complete raid on a *Formica* colony by placing it in a box 2 meters from a *Polyergus* colony and drawing an artificial amazon trail to the edge of the container.

Mary Talbot believed that scouts do not lead their sisters back to the target nest by running at the head of the phalanx. This may not be entirely correct. Although it is true, as she proved, that excited workers of the Michigan species are capable of following odor trails without further guidance, other signals are probably used. Howard Topoff, an entomologist at the American Museum of Natural History, found a more complicated story in another species of amazon ant living in Arizona. During naturally occurring raids, he observed, scout ants always lead the phalanxes. He moved prominent features in the ants' environment, such as bushes and rocks, and proved that these visual cues are more important to the leaders than are chemical trails. After the attack, however, the workers orient homeward both by visually fixing on the guideposts and by following the chemical trails laid by the outward-bound leaders.

Amazon warriors steal the young of other species by rushing into battle and using deadly weapons to slaughter every ant in their way. That might seem the most straightforward, effective way to perform the deed. But more subtle ways to kidnap slaves exist. Wilson, working with Fred Regnier of Purdue University, noticed that another slave-raider of the United States, *Formica subintegra,* is highly successful even without the battlefield élan of the *Polyergus* warriors. The workers possess ordinary mandibles instead of the curved saber-shaped weapons that characterize the amazons, yet they appear to be equally efficient at capturing slaves. Like the amazons, they favor other species of *Formica.* In seeking the key to their success, Wilson and Regnier discovered that each *subintegra* worker possesses a hugely enlarged Dufour's gland that fills nearly half the abdomen. When the raiders attack a colony, they spray "propaganda substances" from this gland on and around the defenders. This material, which is a mix of decyl, dodecyl, and tetradecyl acetates, attracts the *subintegra* raiders, but it alarms and disperses the defenders. The three acetates imitate the true alarm pheromones of the *Formica* victims. They are in effect super-alarm pseudo-pheromones: they are quickly detected by the intended victims, and their odor lingers in the nest long after ordinary alarm substances (such as undecane) have evaporated to levels that are undetectable.

To human observers the captured workers may seem to be slaves; but they themselves act as if they were free. They behave exactly as though they were sisters of the slave-makers, performing the same tasks they would undertake if safe at home with their own colony. This fidelity should come as no surprise. Free-living ants have been programmed by evolution to act in such a way regardless of context, and the slave-makers—programmed in their own, special way—have taken advantage of that instinctual rigidity. Wilson found one slave-maker in Wyoming,

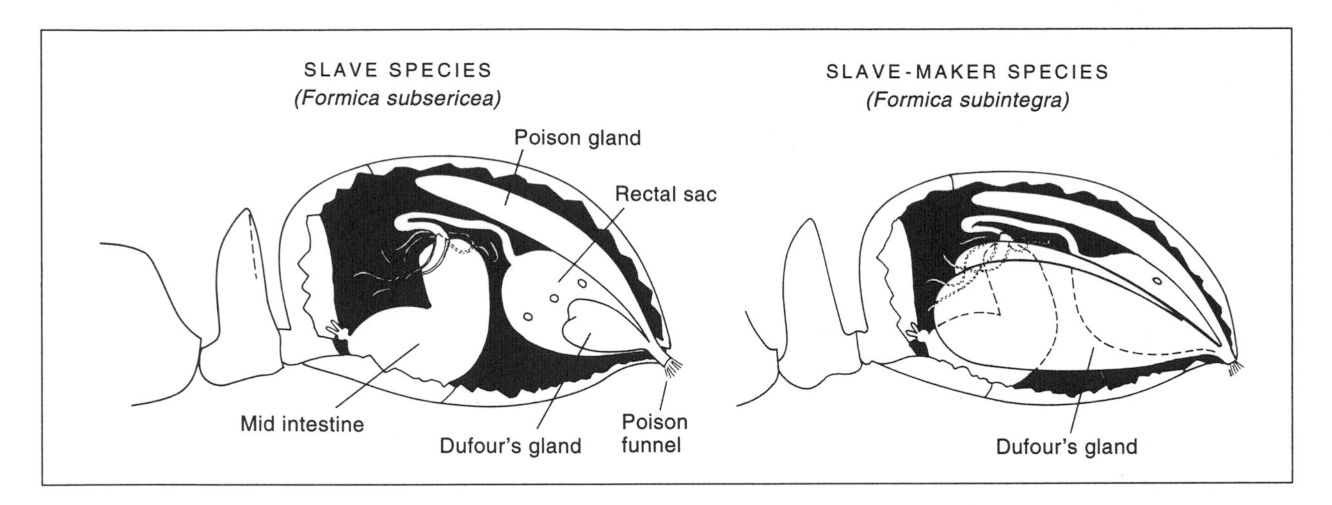

The American slave-making ant *Formica subintegra* uses "propaganda substances" produced in large quantities in its greatly enlarged Dufour's gland. These substances, resembling alarm pheromones, confuse and scatter the defenders. The *subintegra* Dufour's gland is contrasted with the normal-sized organ of *Formica subsericea,* which is not a slave-maker. (From F. E. Regnier and E. O. Wilson, *Science,* 172: 267–269, 1971.)

Formica wheeleri, that employs slaves of more than one species, each with a slightly different program. The outcome is a division of labor resembling a caste system. One of the slave species, *Formica neorufibarbis,* is aggressive and excitable. On a slave raid observed by Wilson, workers accompanied their *Formica wheeleri* mistresses as janissaries. They also helped the *wheeleri* defend the upper reaches of the mixed nest when it was dug open. The other species, a member of the *Formica fusca* group, remained in the lower depths of the nest and attempted to flee and hide when the nest was exposed. Their abdomens were distended with liquid food, and they appeared to have been serving as nurses of the *wheeleri* young.

Several hundred species of ants around the world are known to have become social parasites of other ants, and—at a guess—a few hundred more have the potential to take this evolutionary path. But thousands of mites, silverfish, millipedes, flies, beetles, wasps, and other small creatures have also made the commitment. The ant colony is porous to invasion by this mob of confidence artists by virtue of the simplicity of its communication codes. It is also, from the point of view of the would-be guests, an ecological island lavishly endowed with nutrients and waiting to be exploited. The colony and the nest offer many kinds of niches that predators and symbionts can enter. The exploiters can choose among the foraging trails of the ants, outer nest chambers or guard nests, storage rooms, queen chambers, and nurseries—the last further divided into spaces for pupae, larvae, and eggs.

Or, as often happens, the more brazen guests can live on the bodies of the ants themselves. The extreme of this trend is displayed by certain kinds of mites that ride on army ants in the forests of the American tropics. Tiny, vaguely spiderlike in form, some of the species sit on the heads of the workers and steal food directly out of their mouths. Others lick oily secretions off the ants' bodies, or suck their blood. Apart from food choice, the invading species tend to be highly particular about the part of the body they occupy. Some spend most or all of their time on the mandibles, others on the head, thorax, or abdomen. One whole group, mites of the family Coxequesomidae, fasten themselves exclusively to either the antennae or the coxae, the uppermost segments of the legs. To endure such visitation is rather like having a vampire bat hanging from your ear or a snake wrapped like a garter around your upper thigh.

In our judgment, however, the most extraordinary adaptation of all is exhibited by a macrochelid mite *(Macrocheles rettenmeyeri)* that spends its life sucking blood from the hind feet of the soldier caste of one species of army ant *(Eciton dulcius)*. The mite is nearly as large as an entire segment of the ant's foot. Its presence is analogous to the attachment of a slipper-sized leech to the sole of a human being's foot. Yet in spite of its grossness, the mite does not cripple its host. It allows its entire body to be used by the soldier as an extension of the foot, and the ant walks on the mite without apparent discomfort. That is not all. As observed by Carl Rettenmeyer, the American entomologist who discovered the species, army ants at rest form clusters by hooking the claws of their feet over the legs or other parts of the bodies of other workers. When a *Macrocheles* mite fastens onto a soldier's foot, it lets its hindlegs serve in place of the ant's claws. To accomplish the substitution, the mite bends its legs to just the right curvature and holds them rigidly in position whenever the soldier hooks onto another ant. Rettenmeyer could see no difference in the behavior of ants hanging by their own claws and those suspended from the hindlegs of the parasite.

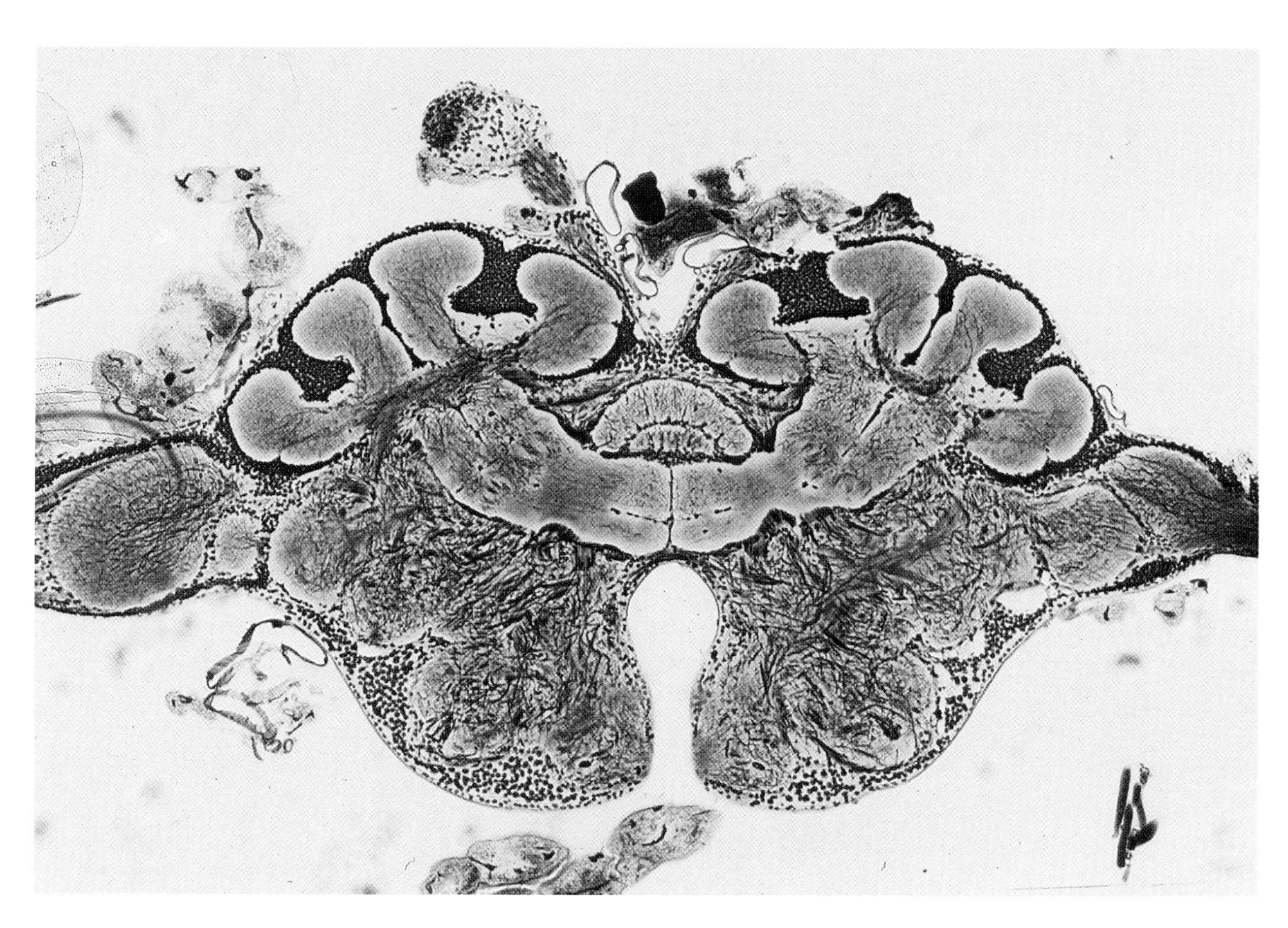

The ant brain is extraordinarily complex for such a tiny organ. This cross-section of the brain of a queen carpenter ant of *Camponotus ligniperda* shows the elaborate "mushroom bodies" at the top, peculiar paired structures consisting of dense masses of nerve cells that process and integrate information. The brain has a structure fine enough to allow ants to learn such simple information as colony odors and the location of several places outside the nest. (Histological preparation and photograph by Malu Obermayer.)

In the first step of vegetation processing to create a fungus garden, a media worker of the leafcutter ant *Atta sexdens* scissors off a section of a leaf outside the nest.

A leafcutter ant of the media caste carries a newly excised section of a leaf back to the nest. Sometimes members of the minor caste ride on the leaf fragment *(below);* their main role appears to be to protect the carrier ant from parasitic phorid flies.

Two media leafcutters cooperate to cut a twig, which will be carried back to the nest and added to the fungus garden.

Inside the leafcutter nest, workers process the vegetative fragments into a garden substrate and cultivate a fluffy white mold on it. The fungus belongs to a species found only in the ant nests.

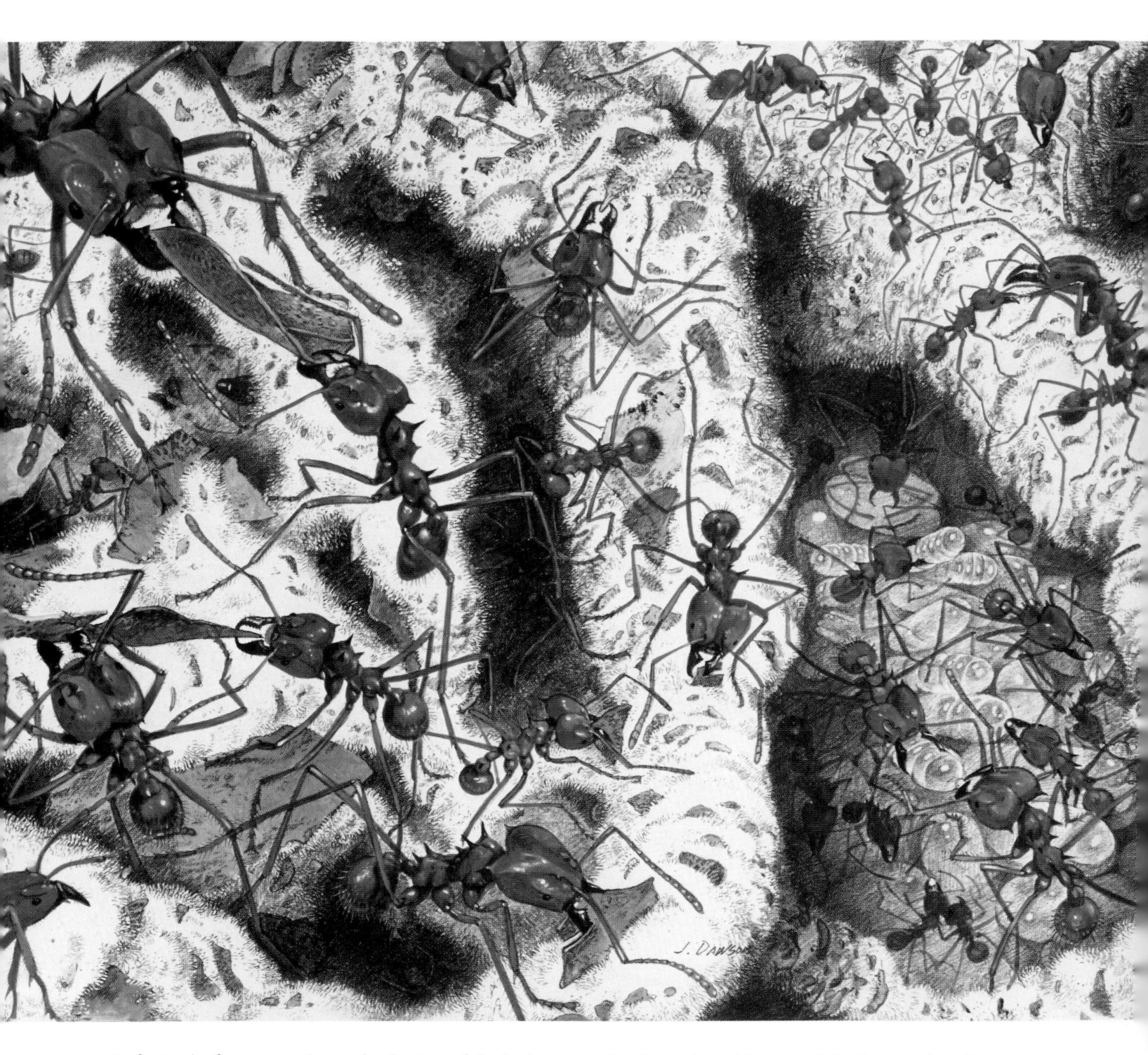

Each step in the preparation and culturing of the leafcutter garden is conducted by a specialized caste of workers, which differ by size and a multiplicity of other anatomical traits. (Painting by John D. Dawson, courtesy of the National Geographic Society.)

The queen of the leafcutter colony, shown here on a section of the garden, is gigantic in comparison with her daughter workers.

When an *Atta* colony has reached a certain size, majors (also called soldiers) are raised. In the picture above we see a soldier pupa surrounded by media sister workers; note the big eye and the large mandibles already visible in this pupal stage. The lower picture shows an adult soldier. The soldiers of the leafcutter colonies are specialized almost entirely for defense. Their sharp jaws, powered by muscles that cram their swollen head capsules, can cut through leather.

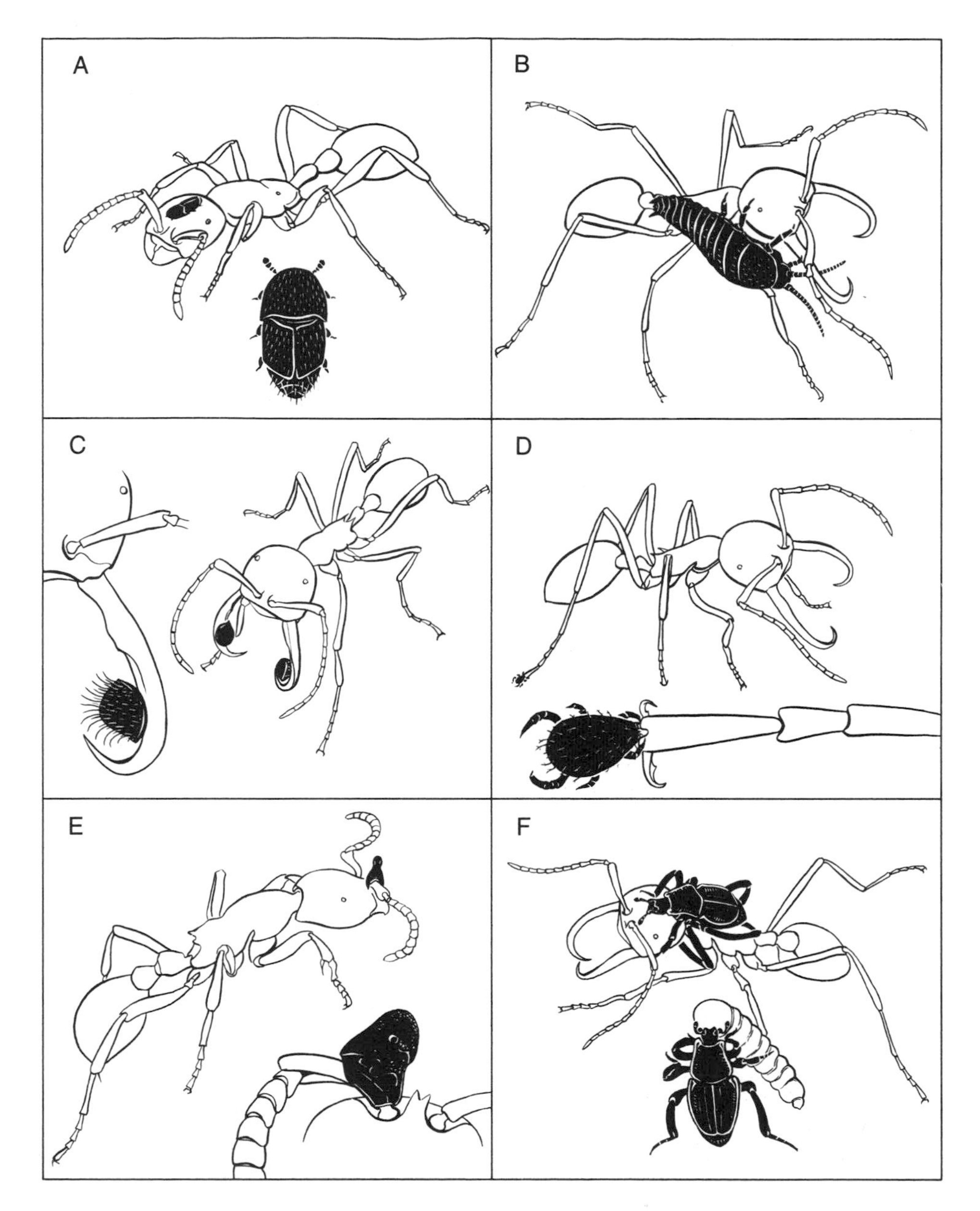

Six arthropod guests (in black) of army ants, which display a few of the many kinds of symbiotic adaptations to these hosts. *(A) Paralimulodes wasmanni* is a limulodid beetle that spends most of its time riding on the body of host workers *(Neivamyrmex nigrescens). (B) Trichatelura manni* is a nicoletiid silverfish that scrapes and licks the body secretions of its hosts (*Eciton* species) and shares their prey. *(C)* The *Circocylliba* mite belongs to a species specialized for riding on the inner surface of the mandibles of major workers of *Eciton. (D) Macrocheles rettenmeyeri* is another mite species which is normally found attached in the position shown, serving as an extra "foot" for workers of *Eciton dulcius.* *(E) Antennequesoma* is a mite genus highly specialized for attachment to the first antennal segment of army ants. *(F)* The histerid beetle *Euxenister caroli* grooms the adults of *Eciton burchelli* and feeds on their larvae. (Drawings by Turid Forsyth.)

The means by which insects and other arthropods trick and rob ants are vast in number. One extreme stratagem, studied by Bert Hölldobler in Germany, is used by the European nitidulid beetle *Amphotis marginata.* This cunning insect, which resembles a small squashed turtle, is the "highwayman" of the local ant world. The beetles hide during the day in shelters along the foraging trails of the shining black ant *Lasius fuliginosus.* At night they patrol back and forth along the trails, occasionally stopping and stealing food from workers running homeward. Ants whose crops are heavily laden with liquid food are easily deceived. The beetles induce them to regurgitate a droplet of liquid food by drumming their short, clublike antennae on the ants' heads and lower lip surfaces, the same signal used by the workers themselves. Soon after the *Amphotis* start to feed, however, the ants realize they have been tricked, and they attack the intruders. The *Amphotis* are in no great danger. They simply retract their legs and antennae under their broad carapaces and flatten themselves on the ground. With the aid of special hairs on their legs they grip the soil surface tightly. The ants are unable to lift the beetles or turn them over. The little highwaymen simply wait for the ants to leave, then amble on down the trail in search of their next victims.

Many kinds of insect predators settle near ant trails for the purpose not of robbing but of killing and eating the passing workers. Yet they seldom use brute force to achieve their ends, because the ants are heavily armed with stings or poison secretions, move in groups, and are quite capable of counterattacking and turning the tables on their tormentors. So the predators employ subtle techniques to capture ants without being recognized and attacked themselves. One method is the kind of wolf-in-sheep's-clothing deception used by the assassin bug *Acanthapsis concinnula.* This insect, whose long wicked-looking beak folds out like the blade of a pocket knife, hunts around the nests of fire ants. It isolates and captures one ant at a time, immobilizes it by injecting poison through the beak, and sucks out its blood. Then it lifts the shriveled corpse onto its own back and sticks it there. The accumulated shield of dead victims provides an excellent disguise and even attracts other fire-ant workers made curious by the presence of their dead nestmates. If ants made motion pictures for entertainment, like people, their favorite horror movie monster would surely be the assassin bug *Acanthapsis concinnula.*

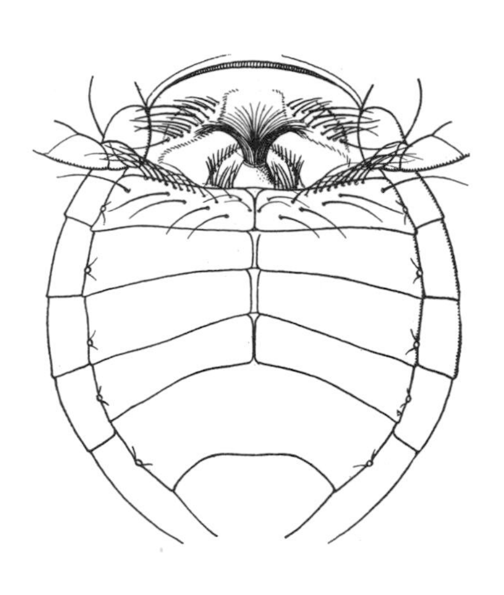

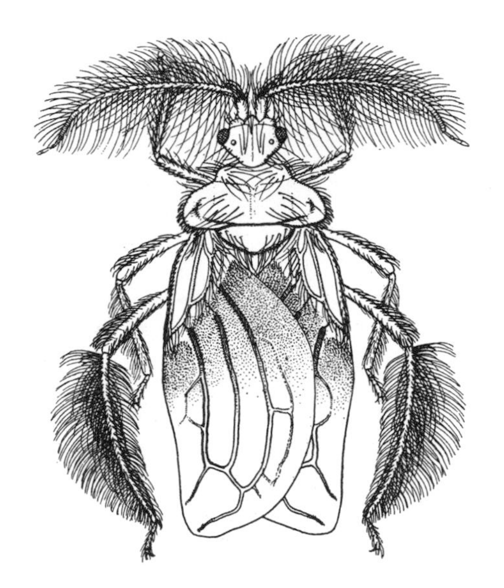

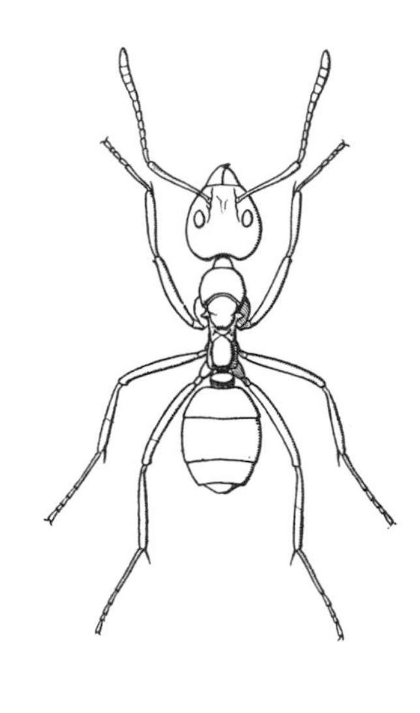

The predatory Indonesian bug *Ptilocerus ochraceus (center)* and the ant *Dolichoderus bituberculatus (right),* on which it feeds. At left is the undersurface of the bug's abdomen with specialized hairs that release a tranquilizer which attracts the ants. (Modified from W. E. China.)

A close rival for this distinction, a kind of dracula of the insect world, is another assassin bug, *Ptilocerus ochraceus,* which feeds on *Dolichoderus bituberculatus,* a species of ants extremely abundant in southeastern Asia. These predators simply sit in or near an ant path, wafting an intoxicating attractant from glands located on the underside of their abdomens. When a worker approaches a *Ptilocerus,* the bug raises itself upon its middle legs and hindlegs and presents the gland surface for closer inspection. The ant moves close and starts to lick the secretions. The bug folds its front legs gently around the body of the ant and brings the tip of the beak into position on the back of the ant's neck. But it does not yet stab the worker or even squeeze it with its legs. The ant continues to feed, and after a few minutes it shows signs of paralysis. When it has become completely helpless, curling itself up and drawing in its legs, the bug pierces it with the beak and sucks out its blood. Thus a *Ptilocerus* is able to kill one worker after another without disturbing the stream of nestmates running close by.

Deceptive chemicals like those employed by the *Ptilocerus* assassin bugs are the stock in trade of a great majority of the more sophisticated predators and social parasites. The prime targets of these invaders are

At the top, a *Formica* worker regurgitates to an *Atemeles* larva. Glands believed to be the source of a false brood identification odor are located in pairs on the dorsal surface of each of the body segments of the *Atemeles* larva. At the bottom, a larva of *Atemeles pubicollis* feeds on a larva of one of its host ants *(Formica)*.

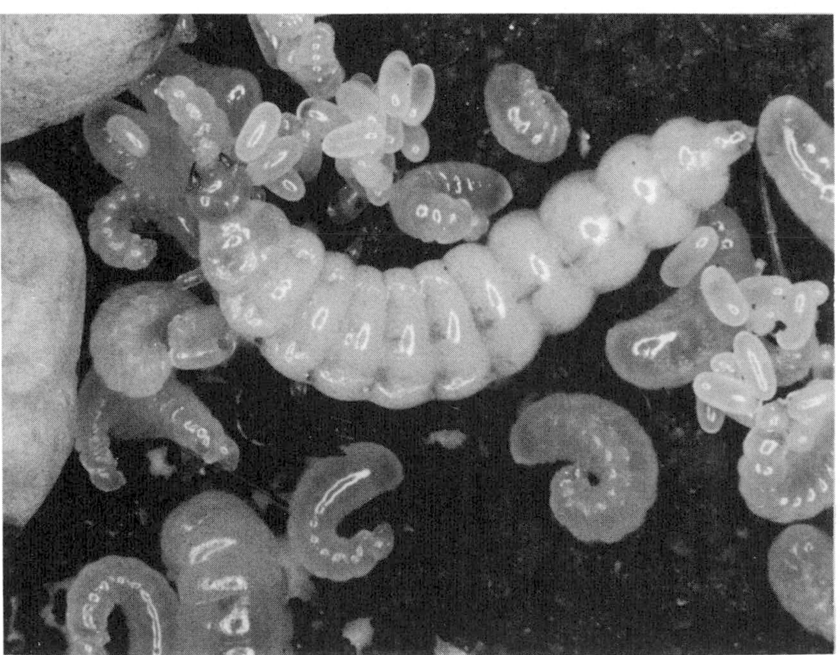

the brood chambers of the ant nest, where the queen and immature stages live. Into this central region are brought the largest quantities of food, and here predators can also find vulnerable piles of fat, helpless larvae and pupae. The brood chambers are nevertheless difficult to penetrate, because they are fiercely defended by the ants. Brood chambers are the combined central headquarters and Fort Knox of the nest. Only animals with very special devices are able to insinuate themselves so deeply, much less survive there for more than a few minutes.

The trick has been accomplished by some of the evolutionarily advanced rove beetles (the family Staphylinidae). Among them, the European species of *Atemeles* and *Lomechusa* are the most skillful known experts. Bert Hölldobler became acquainted with these insects while still a boy through his father's studies, and he learned more from early publications by the famous German priest-entomologist, Erich Wasmann. While still a postdoctoral assistant in Frankfurt he set out to probe as deeply as he could into the lives of the rove beetles. He began by confirming that they live in the nests of *Formica* ants, which are large, aggressive, red and black insects abundant throughout Europe. Some of the beetle species also spend part of the year with ants of the genus *Myrmica,* also abundant in Europe but smaller than the Formicas and more slender in body shape.

A well-known example of such a beetle is *Atemeles pubicollis.* During its larval stage it lives in the nest of the mound-making wood ant *Formica polyctena.* The ants accept the parasite larvae into their brood chambers, and treat them as though they were their own young. Hölldobler was able to prove that the imitation is accomplished by presentation of both mechanical and chemical signals. The beetle larvae beg food by repeating the movements used by the ant grubs themselves. When a passing worker touches the larva, the parasite rears upward to make contact with the ant's head. If successful, it next taps the undersurface of the ant's jaw with its own mouthparts. This sequence is essentially the same as that of the ant larva, except that it is more vigorous. By offering the ant colony liquid food labeled with radioactive materials, Hölldobler was able to measure the rate and direction of flow of the food as the members of the colony subsequently regurgitated it back and forth. He found that the parasite larvae obtain a greater share of the food than do the host ant larvae. In essence they act like cuckoos—birds whose young grow up as parasites in the nests of other species—by persuading the host ants to favor them over their own species. The mistake creates a double burden for the hosts, because the beetles also eat their larvae. The parasites are prevented from destroying the colony altogether only by the fact that they are cannibals: when they grow abundant enough to come in contact, they start to eat one another.

The ant workers also wash the parasites with their moist tongues, using the same motions used to groom their own larvae. Evidently the beetles exude an attractive chemical similar to the one coating the ant larvae themselves. To test this hypothesis, Hölldobler followed a classic experimental procedure for detecting chemical signals. He covered the bodies of freshly killed beetle larvae with shellac to prevent release of secretions. He then placed the corpses outside the nest entrance of a *Formica* colony, and next to them he laid freshly killed but otherwise untreated beetle larvae to serve as controls. The ants quickly carried the control larvae into the brood chamber, as though they were alive and still attractive (recall that ants recognize corpses only from the scent of decomposition products accumulated over several days). The shellac-coated larvae, in contrast, were carried to the garbage dump. If even a small part of the bodies of treated individuals was left free of shellac, they were also carried into the brood chambers. Approaching the problem from another direction, Hölldobler extracted most or all of the secretions from the beetle larvae with solvents. These larvae were no longer attractive. When he added the extract to the leeched individuals, they became attractive again. Finally, when he soaked the extract into paper dummies, these too were carried to the brood chamber. Clearly the identifying trait of ant larvae for their adult caretakers is chemical in nature, and the rove beetles have broken that code.

The *Atemeles* beetles have two homes with ants, one for the summer, the other for winter. After the larvae have pupated and hatched in a *Formica* nest, the adult beetles emigrate in the fall to nests of the ant genus *Myrmica.* The reason for this remarkable shift is that *Myrmica* colonies maintain brood and a food supply throughout the winter, whereas *Formica* ants suspend the rearing of young in that season. In the *Myrmica* nest the beetles, still sexually immature, can feed themselves and

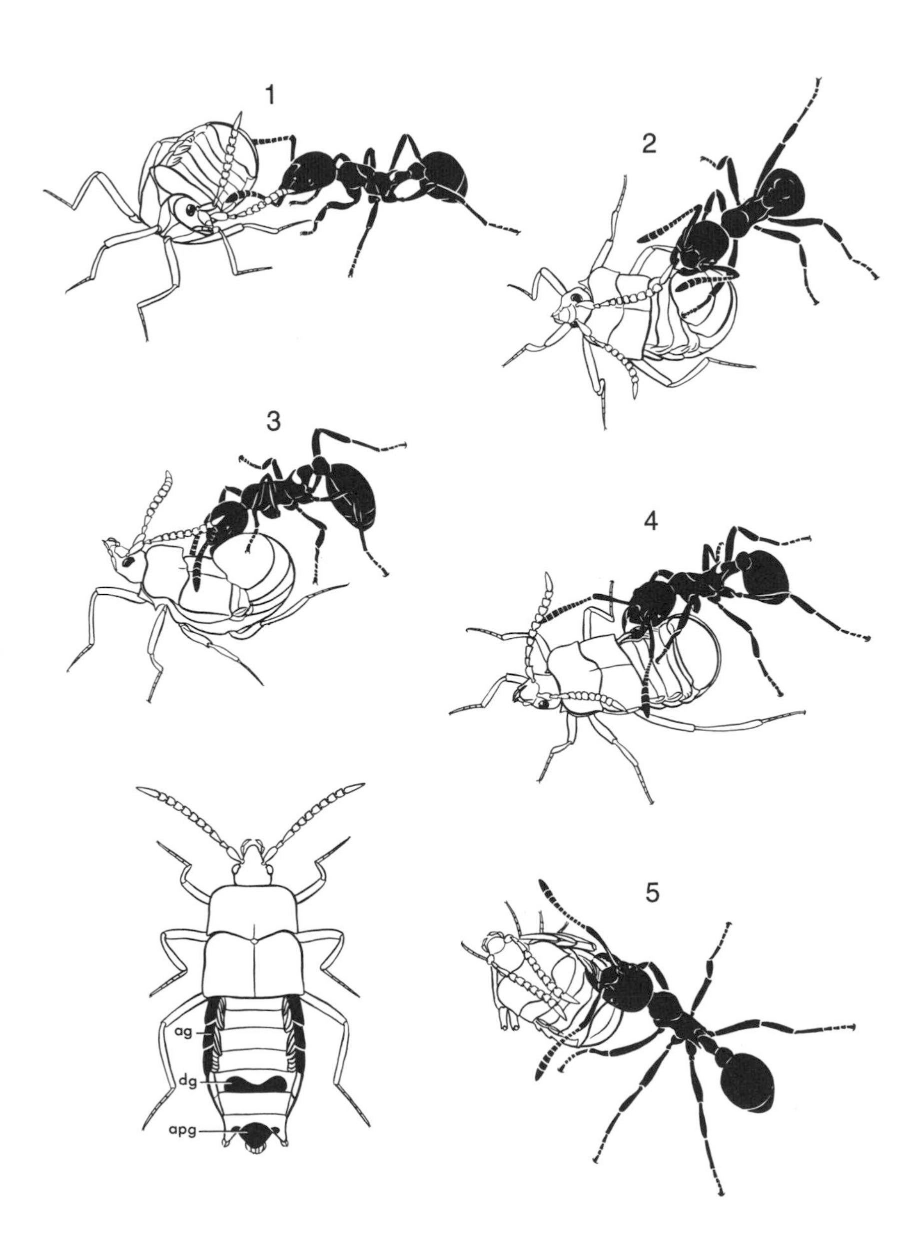

Adoption of the European staphylinid beetle *Atemeles pubicollis* by one of its host ants, a species of *Myrmica*. The drawing at the lower left shows the location of the three principal abdominal glands of the parasite: adoption glands *(ag)*, defensive glands *(dg)*, and appeasement glands *(apg)*. The beetle presents its appeasement gland to a worker of *Myrmica* that has just approached it *(1)*. After licking the gland opening *(2)*, the worker moves around to lick the adoption glands *(3, 4)*, after which she carries the beetle into the nest *(5)*. (Drawing by Turid Forsyth.)

Food-soliciting behavior of the myrmecophilous beetle *Atemeles pubicollis.* The beetle uses its antennae to stroke a host ant worker, which turns toward the beetle *(top)*. The beetle then taps the ant's mouth with its forelegs *(center)*, causing the ant to regurgitate a droplet of liquid food *(bottom)*. (Drawing by Turid Forsyth.)

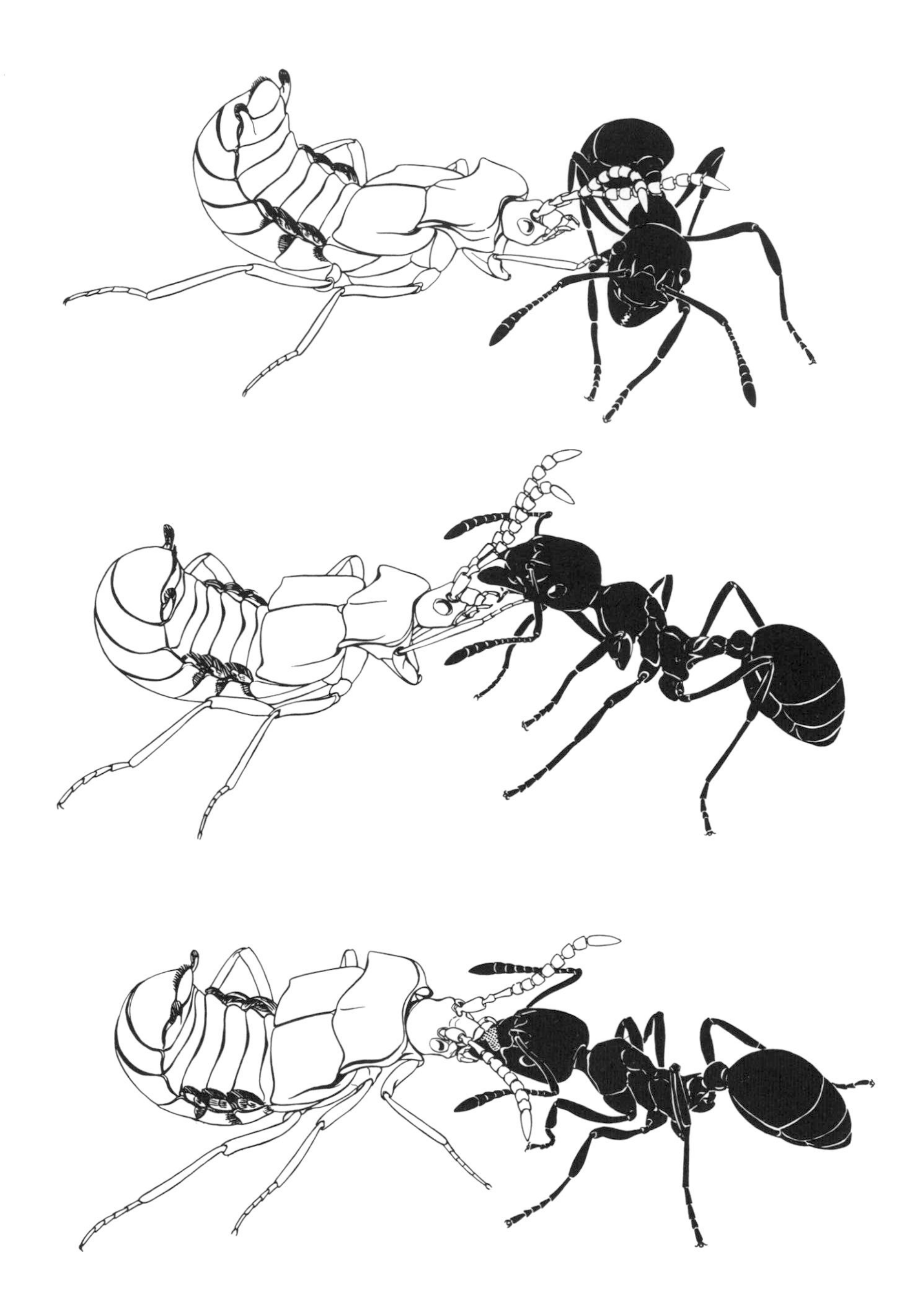

ripen to maturity by the spring, at which time they return to *Formica* nests for mating and the laying of eggs. Thus the life cycles and behavior of the *Atemeles* and the *Formica* and *Myrmica* ants are synchronized in such a way that the beetles can take maximum advantage of the social life of each of the two species that serve as hosts. During these migrations the beetles have to master two tasks. First, they must locate a nest of the alternate host ant each time they move, and second, they must then secure adoption in a potentially hostile environment. To do so they follow four steps in sequence. First, the beetle taps one of the workers lightly with its antennae as though trying to gain the ant's attention. Then it raises the tip of the abdomen and points it at the ant. This region of the body contains appeasement glands, the secretions of which are immediately licked up by the ant and appear to suppress aggressive behavior. The ant then is attracted to a second series of glands located along the sides of the abdomen. The beetle lowers its abdomen in order to permit the ant to approach this part of the body. The glandular openings are surrounded by bristles, which are grasped by the ant and used as handles to carry the beetle into the brood chambers.

By blocking the openings of these glands in rove beetles, Hölldobler found that their secretion is essential for successful adoption. For this reason he called them "adoption glands." Thus the acceptance of the beetles, like that of the larvae, depends on chemical communication and in particular on certain substances that imitate the pheromones secreted by the ants' own young. Inside the nest the beetles live in the brood chambers of their hosts, preying on the ant larvae and pupae. They also solicit food from adult ants by imitating the food-begging signals of the ants.

Propaganda, slavery, decoding, entrapping, mimicry, panhandling, Trojan horses, highwaymen, cuckoos: they are all present among the ants and the predators and social parasites that victimize them. Such words may seem unduly anthropomorphic, turning ants and their associates into little people. But perhaps not. It is equally possible that the number of social arrangements available to evolution anywhere in the world, or even in the universe, is such that the phenomena we have recounted here are inevitable natural categories of exploitation wherever it may occur.

The Trophobionts

EVERYWHERE ants are found, their species have struck a bargain with insects that feed on plants. Aphids, scale insects, mealybugs, treehoppers, and the caterpillars of lycaenid and riodinid butterflies (the latter called "blues" and "metalmarks" in common parlance) give sugary secretions to the ants for food. In return they are protected from enemies. The ants go further, sheltering them with walls of carton or soil, and sometimes they even take them into the nest as virtual members of the colony. This symbiosis, called trophobiosis from the Greek for "nourishing life," has proved one of the most successful in the history of the land ecosystems. It has contributed greatly to the numerical dominance of both the ants and their wards.

The most abundant and familiar trophobionts of the north temperate zone are aphids. Ants and aphids can be found together on the weeds and flowers of almost any garden or weed-strewn field. If you find such an association and watch it for a few minutes, you will see a worker approach an aphid and touch it lightly with her antennae or forelegs. The aphid responds by extruding a drop of sugary liquid from its anus. The ant responds by quickly licking up this honeydew—the term entomologists euphemistically use for aphid excrement. She passes from aphid to aphid, soliciting each one the same way, until her abdomen swells with the accumulated harvest. Then she returns to the nest to regurgitate some of the sweet fluid to her nestmates.

These droplets prized by the ants are not only attractive but highly nutritious. When aphids feed on the phloem sap of plants, drawing it up through their needlelike proboscises through a combination of sap pressure and the pumping of cibarial muscles, they acquire a complete set of the food substances they need. But they do not use all the materials collected this way. Some of the nutrients, including sugars, free amino acids, proteins, minerals, and vitamins, are passed through the gut and out the anus as part of the waste material. In the course of the passage the liquid changes chemically: some of its components are absorbed, while others are converted to new compounds, and still others are added afresh by the aphids from their

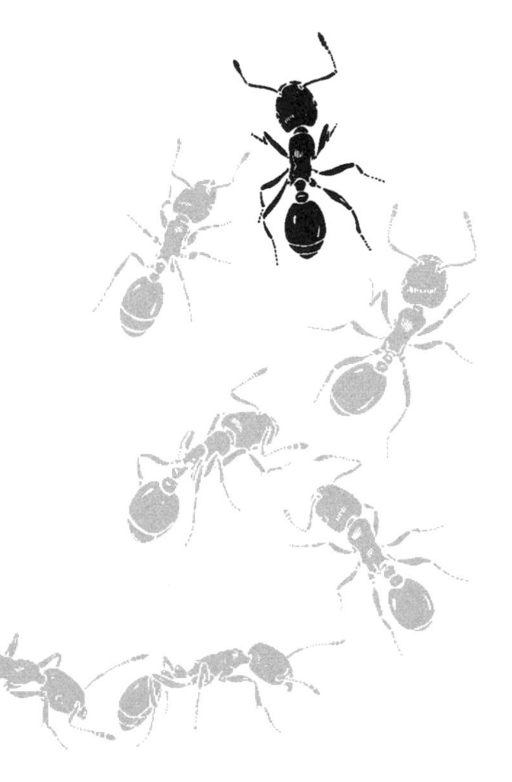

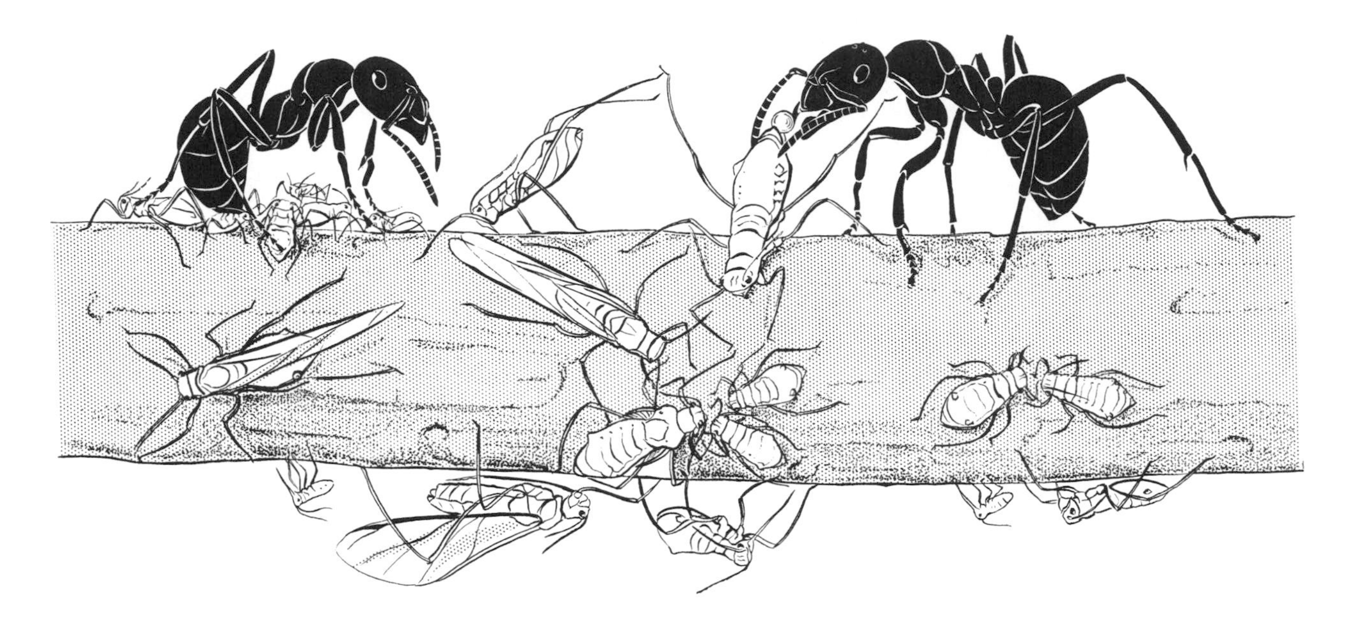

Workers of the European wood ant *Formica polyctena* attending aphids *(Lachnus robaris).*
(Drawing by Turid Forsyth.)

own tissues. Measurements made on one species, *Tuberolachnus salignus,* disclosed that as much as half the free amino acids are absorbed by the aphid's gut, and half passed on. In a few cases the honeydew of aphids contains amino acids that are not present in the sap; they are evidently given to the ants as chemically new metabolic products.

From 90 to 95 percent of the dry weight of honeydew is composed of sugars, most of which are sweet to human taste. The diverse mixtures of sugars passed out in the honeydew, in blends and concentrations that are peculiar to each species of aphid, comprise various amounts of fructose, glucose, saccharose, trehalose, and higher oligosaccharides. Trehalose, the natural blood sugar of insects, is present as 35 percent of the total sugar amount in typical honeydew. The sugars also include two trisaccharides, fructomaltose and melezitose, with the latter making up 40–50 percent of the total sugar. In addition to these and minor amounts of other sugars, the honeydew contains organic acids, B vitamins, and minerals.

Similar bounties of nutrients are provided by other groups of sap-feeding members of the insect order Homoptera, including scale insects (members of the family Coccidae), mealybugs (Pseudococcidae), jumping plant lice (Chermidae), treehoppers (Membracidae), leafhoppers (Cicadellidae), froghoppers, also called spittle insects (Cercopidae), and lanternflies (Fulgoridae). Many of these insects are accessible, easy to exploit, and therefore heavily attended by the ubiquitous ants. In New Guinea one day, while waiting on the edge of a road for a car ride, Wilson succeeded in "milking" giant scale insects surrounded by ants, simply by touching them with hairs from his head in imitation of the requisite antennal tapping by ants. He found the liquid to be detectably sweet. (Such are the informative pleasures that fill the idle hours of naturalists in the field.)

The honeydew yielded by the homopterous insects is a bounty easily available to the ants wherever they can crawl up onto vegetation or search beneath it. Much of the material is simply discarded as ordinary waste. But whether used or not, the output by homopterous insects around the world is prodigious. Aphids in the genus *Tuberolachnus* excrete approximately 7 droplets in an hour, an amount which exceeds their own body weight. Sometimes the honeydew accumulates in large enough quantities to be usable even by human beings. The manna "given" to the Israelites in the Old Testament account was almost certainly the excretion of the scale insect *Trabutina mannipara,* which feeds on tamarisk. The Arabs still gather the material, which they call man. In Australia, chermid honeydew, or sugar-lerp, is collected as food by the Aborigines. Up to 3 pounds can be collected by one person in a single day. It is a fact not widely appreciated that most honey consumed by people around the world is honeydew collected by bees from the surfaces of bushes and trees. One of our favorite foods is insect excrement processed in the gut of other insects. Not surprisingly, then, ants also gather honeydew of all kinds, in large quantities and under a wide range of circumstances. Many, perhaps most, species collect it from the ground and vegetation wherever the liquid happens to fall. But it is only a short step in evolution for the ants to solicit the honeydew directly from the homopterans themselves.

The easy reciprocity of the symbiosis has driven many ants and their trophobionts during the course of their evolution to extremes of adaptation. A few ant species, to be described shortly, have become completely dependent on their partners, tending them like domestic cattle. For their part many trophobionts have acquired structural and behavioral adaptations for life with ants. Aphids frequently associated with ants tend to be less able to repel enemies. They possess small cornicles, the hornlike projections at the rear of the abdomen from which noxious chemicals are secreted. They also have a thinner protective coating of wax on their bodies than do aphids untended by ants. The job of defense clearly has been ceded to their formidable partners.

Aphid species not dependent on an ant association forcibly propel the honeydew droplets well away from the body. This hygienic measure keeps them free of the gummy liquid and the fungi that flourish on such material. In contrast, the trophobiont aphids make no effort to get rid of their honeydew, but present the material in a manner that lets the ants feed efficiently. They ease out droplets one at a time and hold them for a while on the tips of their abdomens, just outside the anus. Many species possess a basket of hairs that holds the honeydew firmly in place. If a droplet is not accepted by the worker ants, the aphid often draws it back into the abdomen, to be offered at a later time.

Honeydew has thus been converted in the course of evolution from mere excrement into valuable barter. What do the trophobionts receive for this service to the ants? The primary answer is the superb defense force provided by their hosts. The ants drive off the parasitic wasps and flies that would otherwise inject eggs into the aphids' bodies. They also drive away the lacewing larvae, beetles, and other predators that prowl the vegetation and slaughter unprotected homopterans like wolves loosed on flocks of sheep. The trophobiont herds grow large and densely packed under the ants' protection. In some cases their caretakers move them from one place to another to provide better protection or fresher food sources.

The eggs of the American corn-root aphid, for example, are kept by colonies of the ant *Lasius neoniger* in their nests throughout the winter. The following spring the workers transport the newly hatched nymphs to the roots of nearby food plants. If the plants die, the ants move the aphids to other, undisturbed root systems. During the late spring and summer, some of the aphids grow wings and fly away in search of new plants. After they land and start to feed, they may be adopted by other ant colonies in whose territories they happen to have settled. The *Lasius* workers incorporate their guests into the colony in the fullest sense. They mix the aphid eggs with their own. And when they emigrate to a new nest site, they pick up the eggs—or in the warm season, the nymphs and adults—and transport them gently and unharmed to the new location. At all times they protect the aphids from invaders with the same devotion they give their own young.

The ants do not respond the same way to all trophobionts they identify as nestmates. Some of their behavior appears designed to cater specifically to the needs of their guests. They carry the insects not only to a plant on which the trophobionts might feed, but to the correct species of plant—and, more precisely, to the part of the plant appropriate to the correct stage of the insects' development.

Even more impressively, the queens of a few ant species carry scale insects in their mandibles when they depart from the nest on their nuptial flights. After mating and settling to the ground, they are ready to start a new colony with a mother trophobiont in place to provide honeydew. This activity, comparable to human homesteading with a pregnant cow in tow, has been witnessed in one species of *Cladomyrma* in Sumatra and in several species of *Acropyga* in China, Europe, and South America. It is likely that the behavior will be discovered in yet other kinds of ants.

In at least one case, the trophobionts assist in their own transport by hitching a ride. The behavior has been observed in small teardrop-shaped mealybugs of the genus *Hippeococcus* in Java that live as guests in the underground nests of *Dolichoderus* ants. Under ant guardianship the homopterans feed on the branches of nearby trees and shrubs. When the nest or harvesting site is disturbed, many of the bugs are picked up and carried away by the worker ants in the usual manner. But others climb onto the bodies of their hosts and ride to safety. Ant riding is facilitated by the long, grasping legs and the suckerlike feet that distinguish the *Hippeococcus* mealybugs.

Some species of ants have become wholly dependent on their insect

Javan mealybugs of the genus *Hippeococcus* escape from danger by climbing onto the backs of their host ants and allowing themselves to be carried to safety. Their legs and tarsi are apparently specially modified for this purpose. Here three individuals are shown being carried by a worker of *Dolichoderus*. (Drawing by Turid Forsyth.)

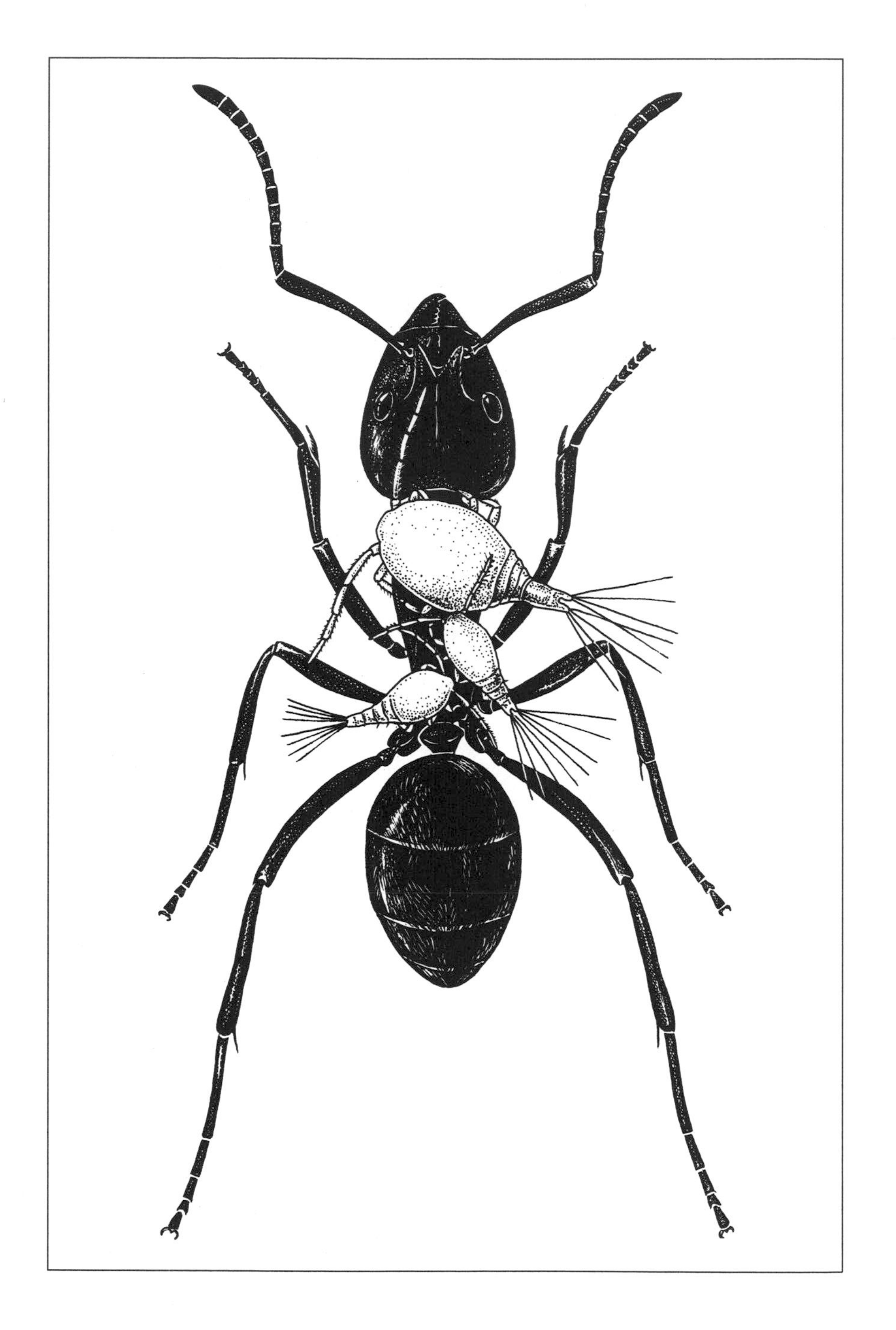

cattle. This ultimate level of specialization appears to have been attained by small-eyed, subterranean ants of the genus *Acanthomyops*, which are abundant through the cold temperate regions of North America. It also appears to characterize the physically similar species of the genus *Acropyga*, which are found in tropical and warm temperate regions around the world. Honeydew may form the sole diet of these ants, which tend herds of mealybugs and other homopterans on the roots of plants. But it is also possible that the ants obtain extra protein by eating some of the insects. Such herd thinning of scale insects has been observed in African weaver ants. When presented with excessive numbers of the trophobionts in experiments, weaver-ant workers were observed to kill individuals until the population reached the level needed for a sufficient but not excessive flow of honeydew.

The most complete and remarkable trophobiosis of all was discovered during the early 1980s, by Ulrich Maschwitz and his collaborators in Malaysia. It consists of a way of life never before encountered in ants: true nomadism, or full migratory herding. The ant colonies live as stock farmers. They subsist entirely on their herds and closely coordinate their lifestyle with that of the livestock, while accompanying them from one pasture to the next.

The ants are *Dolichoderus cuspidatus* and several other species of the same genus, dwellers in the rain forest canopy and shrub understory, and the "cattle" are mealybugs of the genus *Malaicoccus*. The mealybugs feed entirely on the phloem sap of trees and shrubs of the forest. They are carried by the ants to the feeding sites, some of which are more than 20 meters from the nests. The nests are located in the dense vegetation between leaves or in preformed cavities in wood. The workers do little or no conventional house building on their own. Instead, they form the walls and cavities of their inner domicile solely from their own bodies, very much like army ants. They cling to one another to create a solid mass that shields the brood and mealybugs.

The trophobionts are treated as full members in the herder colony. The adult females are often mixed with the larvae and other immature forms of the ants. They are viviparous, giving live birth to their young while in the secure heart of the worker mass. A mature colony of the nomadic *Dolichoderus* contains a single queen, over 10,000 worker ants, about 4,000 larvae and pupae, and more than 5,000 mealybugs. The nests and feeding sites are connected by heavily traveled odor trails. The transport of the trophobionts back and forth between the two sites is intense; at any given time 10 percent of the workers running on the trails carry mealybugs between their mandibles. Since the young, sap-laden plant shoots preferred by the bugs are quickly exhausted, the ants frequently have to locate new feeding sites and transfer the grazing herds to them.

When the distance between the nest and the feeding site becomes too large for easy transport, the *Dolichoderus* colony simply moves en masse to the feeding site. During the emigration the brood and mealybugs are carried in a well-organized manner, parked at intervals in depots scattered along the odor trail, then moved along until the entire colony is settled at its final destination. The move can be triggered not only by hunger but also by the physical disturbance of the bivouac or a change in the surrounding temperature or humidity. There is no regularity in the timing of the emigrations. In the colonies studied by Maschwitz and Heinz Hänel during one 15-week period, the frequency varied from two a week to none at all.

At the feeding sites the mealybugs are always attended by *Dolichoderus* workers, which continually harvest the honeydew droplets extruded from the anus of the homopterans. So intense is the activity that the little mealybugs are almost permanently covered by a layer of feeding ants. They emit droplets from time to time and hold them in long bristles on their bodies in a position that allows the fluid to be licked up by the ants. The emission is spontaneous: unlike less specialized trophobionts, the *Malaicoccus* mealybugs do not wait to feel the drumming on the body by the ant's antennae before offering their honeydew.

When the feeding aggregations are disturbed, both the herder ants and the mealybugs begin to move about excitedly. Single mealybugs crawl on top of workers. There they are plucked off by the ants, who carry them out of harm's way. Whereas small mealybugs are merely scooped up where they stand or walk about, the larger ones raise their bodies in a pose that clearly invites the ants to pick them up. During transport the mealybugs remain motionless, except to caress the heads of the ants with light movements of their antennae.

Maschwitz and Hänel believe that the *Dolichoderus* herders never kill mealybugs for meat. They also found no evidence that the workers search for insect prey away from the nest. These ants appear to depend entirely on the honeydew of their trophobiotic partners. When deprived of mealybugs, the colonies decline rapidly. Correspondingly, the *Malaicoccus* herds soon perish if separated from their ant partners. When Maschwitz offered them to other kinds of ants as potential trophobiont guests, they were attacked and carried into the nests as prey. In short, the symbiosis between the nomadic ants and the mealybug herds is a total and indissoluble union.

The proffered gift of ant protection is so generous and ubiquitous as to constitute an open door for evolutionary opportunism. At first the option might seem to be limited to insects that feed on plant sap, creatures that can easily contribute some of the harvested liquid to the ants in the form of sugary excrement. If that were true, insects that prefer plant tissues instead of sap and therefore pass cellulose-laden feces might never be able to offer a nutritious meal to the ants as their part of the bargain. However, another, less direct way to achieve the same end exists. It has been adopted by caterpillars of certain lycaenid and riodinid butterflies, which eat plant tissue but then use some of the nutrients and energy to manufacture honeydew in special glands. The glands are of two known kinds. Scattered across the surface of the caterpillars are perforated structures called pore cupolas, which secrete substances apparently attractive to ant workers. And on the backs of the caterpillars, near the rearmost part of the body, is found Newcomer's gland, called the honey gland by some writers, which secretes a sweetish fluid imbibed by ants. In one European species, the Provence Chalk-hill Blue *(Lysandra hispana)*, the secretion contains quantities of fructose, sucrose, trehalose, and glucose, as well as minor quantities of protein and a single amino acid, methionine. The Australian species *Jalmenus evagoras* also produces a medley of sugars and at least 14 free amino acids, among which the dominant compound, serine, occurs at a much higher concentration than exists in the nectar-producing organs of plants.

Workers of the ant species *Formica fusca* tend final-instar larvae of the lycaenid butterfly *Glaucopsyche lygdamus*. The worker shown above is feeding on liquid from the larva's honey gland. The worker below defends a larva by seizing an attacking parasitic wasp in her mandibles. (Photographs by Naomi Pierce.)

The caterpillars of the lycaenid butterflies thus offer something approaching a balanced diet to the ants caring for them. They are attractive to the workers solely by virtue of the smell and taste of their secretions. In return, the ants protect the caterpillars from their enemies, including ants and predatory wasps that eat them and parasitic flies and wasps that lay eggs on and inside their bodies. In one experiment conducted in Colorado by Naomi Pierce and her collaborators, the adaptive edge provided by the symbiosis was revealed in striking fashion. When groups of larvae of the Silvery Blue *(Glaucopsyche lygdamus)* were experimentally isolated from their attending ants in the field, they survived at a rate of only 10 to 25 percent that of caterpillars nearby allowed to keep their ants.

The advantage of associating with ants is great enough to form a driving selection force in the evolution of the butterflies. The adult females of many lycaenid species search for plants that have particular ant species on them before depositing their eggs, thus ensuring that their young will receive protection from the beginning. This move is sometimes a virtual necessity. Pierce and her colleagues discovered that mortality in the Australian lycaenid *Jalmenus evagoras* caused by predators and parasites is so severe that larvae and pupae left unattended by ants have virtually no chance at all to survive. In addition to providing protection, the ants shorten the time required for the caterpillars to develop, thereby reducing the period that larvae are exposed to threats from their enemies. The association is not entirely cost-free, however. The energy surrendered by the caterpillars in their sugary secretions is high enough to reduce the size they attain as adult butterflies, and adult size in turn is important for attracting mates and increasing the fecundity of the females. Yet the survival imperative of ant protection is so great as to have outweighed these disadvantages in the course of evolution. As a consequence the butterflies have embraced trophobiont symbiosis decisively.

The food obtained by the ants from the lycaenid caterpillars is no mere casual supplement. In Germany, Konrad Fiedler and Ulrich Maschwitz set out to measure the contribution of caterpillars of the blue *Polyommatus coridon,* which is attended by *Tetramorium caespitum,* the

The adoption procedure of the third-instar caterpillar of the Large Blue *Maculinea arion,* a European butterfly. The individual above waits for a host ant and still has the typical shape of a lycaenid larva. The caterpillar at the bottom has been milked by a *Myrmica* worker, and after hunching its body allows itself to be transported back to the nest by the ant. (Drawing by Turid Forsyth.)

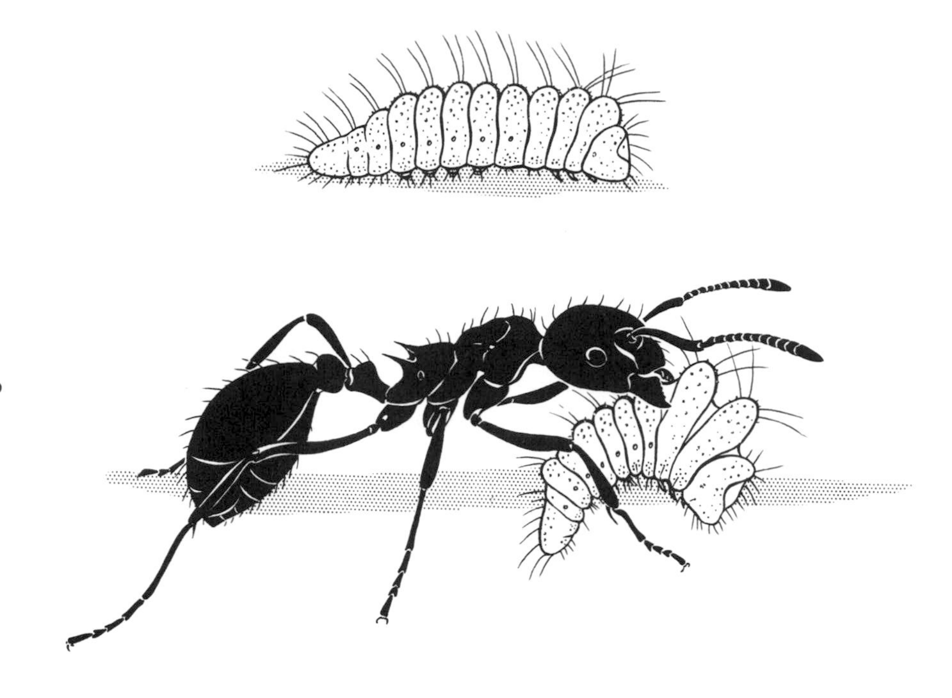

European pavement ant (a frequent pest in U.S. households). They found that a typical population of the caterpillars can deliver from 70 to 140 milligrams of sugar per square meter of vegetation each month, containing a chemical energy of 1.1 to 2.2 kilojoules. This amount is sufficient to cover the entire needs of a small colony of ants if the workers do no more than harvest caterpillar honeydew over an area of 10 square meters.

It is a general rule of evolution that a good thing, in this particular case mutual symbiosis, eventually gets abused by one species or another. Some of the lycaenid butterfly species have evolved a diabolical stratagem to trick and exploit the ants that help them. They not only accept protection but also eat the young of the ants. Such a parasitic form is the Large Blue *(Maculinea arion)* of northern Europe and Asia. Its caterpillar feeds on wild thyme until it reaches its last larval instar. Then it crawls down onto the ground and hides in crevices under tufts of grass until a worker of the common ant *Myrmica sabuleti* finds it. The caterpillar is intensively antennated by the ant and responds by releasing a secretion from its nectar organ. The larva then grotesquely deforms its body. It pulls in its head, swells its thoracic segments, and constricts its abdominal segments, giving its body a hunched, tapered look. Apparently this radical new shape serves as a signal to the worker, which may or may not work in concert with attractive secretory substances on the body of the caterpillar.

Whatever the exact nature of the key stimuli, which remains to be determined by biologists, the ant now picks up the caterpillar and carries it into the nest. Once installed in the nursery chambers of its host, the caterpillar overwinters. In the spring it turns into a carnivore, feeding heavily on the ant larvae. Upon reaching maturity, it pupates in the nest. Finally, in July, it emerges as a winged butterfly to restart the cycle.

Depredations by greedy lycaenid butterflies do not stop with simple predation. Some species intrude on the symbiosis between ants and aphids, scale insects, and other homopteran insects. Species of blues in the genus *Allotinus,* which are common butterflies through much of tropical Asia, exploit this symbiosis in two ways. The adult butterflies alight among the homopterans and feed on the honeydew, then lay their eggs nearby. When the caterpillars hatch, they prey on the homopterans and imbibe their honeydew. The caterpillars apparently do not offer the ants anything in return for these liberties, yet somehow, perhaps through appeasement or false identification substances secreted from their glands, they remain immune from attack.

Army Ants

DAWN BREAKS at the Río Sarapiqui of Costa Rica. As the first light suffuses the heavily shaded floor of the rain forest, there is no trace of a breeze to stir the moist and pleasantly cool air. The hour is announced by the flutelike calls of pigeons and oropendolas perched out of sight in the canopy, punctuated by the distant coughs and roaring of howler monkeys. The treetop inhabitants, first to sense the light, call in the change to the diurnal fauna. The night animals soon fall inactive, and a new cast moves onto center stage.

Beneath the slant of a fallen tree, where the base of the trunk is propped above the ground by thick protruding buttresses, a colony of army ants begins to stir. They are swarm raiders, *Eciton burchelli,* one of the most conspicuous ants in tropical forests from Mexico to Paraguay. The swarm raiders do not build nests like most other ants. They dwell in what Theodore Schneirla and Carl Rettenmeyer, pioneers of the study of army-ant behavior, first called bivouacs, temporary camps in partly sheltered locations. Most of the cover for the queen and immature forms is provided by the bodies of the workers themselves. When the workers gather to establish the bivouac, they link their legs and bodies together with strong hooked claws at the tips of their feet. The chains and nets they form accumulate layer upon interlocking layer until finally the entire worker force constitutes a solid cylindrical or ellipsoidal mass about a meter across. For this reason Schneirla and Rettenmeyer spoke of the resting ant swarm itself as the bivouac.

A half million workers constitute the bivouac, a kilogram of ant flesh. Toward the center of their mass are collected thousands of white larvae and a single heavy-bodied mother queen. For a brief interval in the dry season, a thousand or so males and several virgin queens will be briefly added, but none are present on this and most other occasions.

When the light level around the ants exceeds 0.5 lux, the living cylinder begins to dissolve. Close up, the dark brown conglomerate exudes more of its musky, somewhat fetid odor. The chains and clusters break up and tumble into a churning mass on the ground.

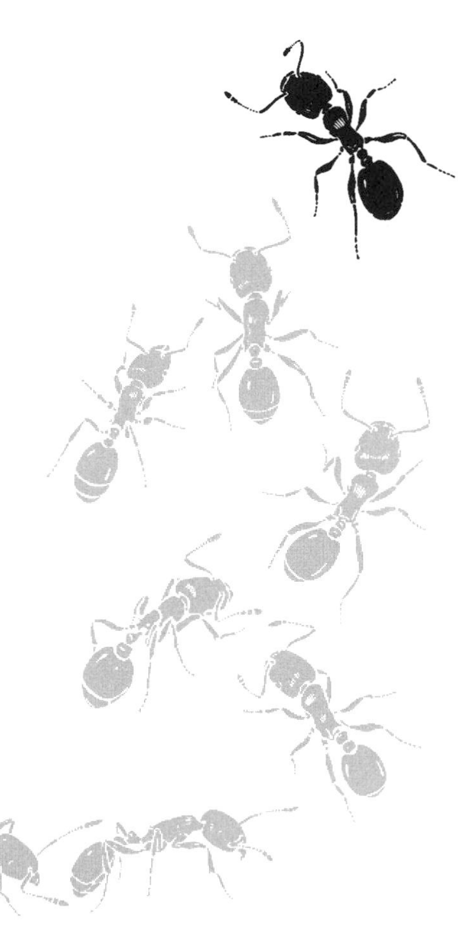

Workers of the army ant *Eciton burchelli* of tropical America join together to form shelters out of their own bodies. First several ants choose a log or some other object near the ground with a space beneath it, then hang from its lower surface with their tarsal claws interlocked. Other ants run down the strands and add their own bodies to create thick ropes that eventually combine into a large mass, the bivouac. (Drawing by John D. Dawson, courtesy of the National Geographic Society.)

As pressure builds, the mass flows outward in all directions, like a viscous liquid poured from a beaker. Soon a raiding column emerges along the path of least resistance and grows away from the bivouac. The tip advances at 20 meters an hour. No leaders take command of the raiding column; any ant can run point. Workers reaching the van press forward alone for a few centimeters and then wheel back into the throng behind them. They are replaced immediately by others who extend the march a little farther. As the workers run onto new ground, they lay down small quantities of trail substances from the tips of their abdomens. These secretions, which originate in the pygidial gland and hindgut, guide others forward. Workers encountering prey deposit extra recruitment trails that draw large numbers of nestmates in that direction. The total effect is to create a swarm whose edge is a broad kaleidoscope of eddies and clumps.

A loose organization also emerges in the rear columns. They are automatically generated from differences in the behavior of the several castes. The smaller and medium-sized workers race along the chemical trails and extend them at the points, while the larger, clumsier soldiers, unable to keep a secure footing among their nestmates, tend to travel on either side. The flanking position of the soldiers misled early observers into concluding that they are the leaders of the army. As Thomas Belt tried to explain in his 1874 classic *The Naturalist in Nicaragua,* "Here and there one of the light-colored officers moves backwards and forwards directing the columns." Actually the soldiers have no visible control over their nestmates. With their large bodies and long, sickle-shaped mandibles, they serve instead almost exclusively as a defense force. The small and medium-sized workers, with shorter, clamp-shaped mandibles, are the generalists. These "minor" and "media" workers, as entomologists call them, are in charge of the quotidian work and the movements of the colony. They capture and transport the prey, choose the bivouac sites, and care for the brood and queen.

The middle-sized swarm raiders also form teams to carry large prey back to the nest. When a grasshopper, tarantula, or other animal is killed that proves too bulky for a single worker to handle, a group of workers gather around it. First one and then the other tries to move it; sometimes

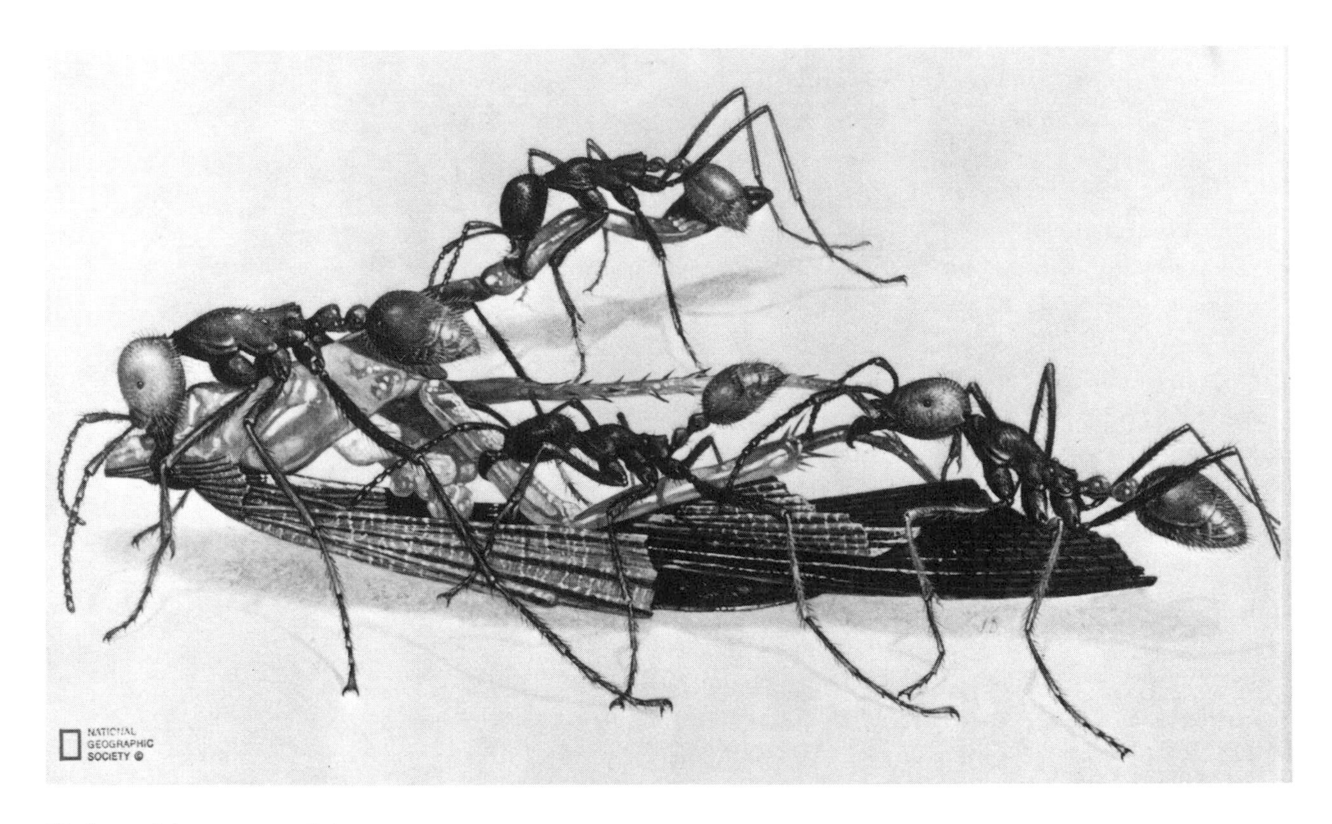

Workers of the army ant *Eciton burchelli* form into teams during the group transport of prey. Here a single large submajor, a member of the caste specialized for this function, is assisted by smaller minor workers in carrying part of the carcass of a cockroach. (Drawing by John D. Dawson, courtesy of the National Geographic Society.)

two or three join forces in tugging at it. One of the largest of the ants, usually a "submajor"—a member of the size class just below that of the fully developed soldier—may be able to drag or carry the prey. Alternatively, the workers cut it into pieces small enough to be handled by a submajor. As the big ant moves the carcass along, smaller nestmates—mostly minors—rush in to help lift and carry it. Now the prey speeds on its way to the bivouac. Nigel Franks, the British entomologist who discovered this behavior, used measurements in the field to show that the teams of the army ants are "superefficient." They can carry items that are so large that if they were fragmented still more, the original members of the group would be unable to carry all the fragments. This surprising result is explained at least in part by the ability of teams to overcome rotational forces, which twist objects away to the side and out of the control of the running ants. Individuals that line up all around the prey while running in the same direction are able to support an object so that the rotational forces are automatically balanced and largely disappear.

Eciton burchelli has an unusual mode of hunting even for an army ant. The armies of this swarm raider do not run in narrow columns but spread out into flat, fan-shaped masses with broad fronts. Most other army-ant species (as many as ten or more may coexist in the same tract of tropical forest) are column raiders, pressing outward along narrow trails in columns that split and rejoin and split again to form treelike patterns in their search for prey.

If you wish to find a colony of swarm raiders in Central or South America, an experience well worth the effort, the quickest way is to walk quietly and slowly through a tropical forest in the middle of the morning—just listening. For long intervals the only sounds you are likely to hear are birds and insects in the distance, mostly in the understory and in the crowns of the higher trees. Then comes the "chirring, twittering, and piping" of antbirds, as one observer put it. These are the specialized thrushes and wrenlike forms that follow the *Eciton burchelli* raids close to the ground in order to feed on insects flushed out by the marching workers. Then you will hear the buzzing of parasitic flies that hover and dart in the air above the swarms, occasionally dive-bombing to deposit an egg on the backs of the escaping prey. Next comes the murmur and hiss of the countless prey themselves, running, hopping, or flying out ahead of the advancing ants. Drawing closer to the action, you may catch a glimpse of ant butterflies, narrow-winged ithomiines that fly over the leading edge of the swarm and stop at intervals to feed on the droppings of the antbirds.

Close behind the victims and hangers-on are the destroyers themselves. "For an *Eciton burchelli* raid nearing the height of its development in swarming," Schneirla wrote, "picture a rectangular body of 15 meters or more in width and 1 to 2 meters in depth, made up of many tens of thousands of scurrying reddish-black individuals, which as a mass manages to move broadside ahead in a fairly direct path. When it starts to develop at dawn, the foray at first has no particular direction, but in the course of time one section acquires a direction through a more rapid advance of its members and soon drains in the other radial expansions.

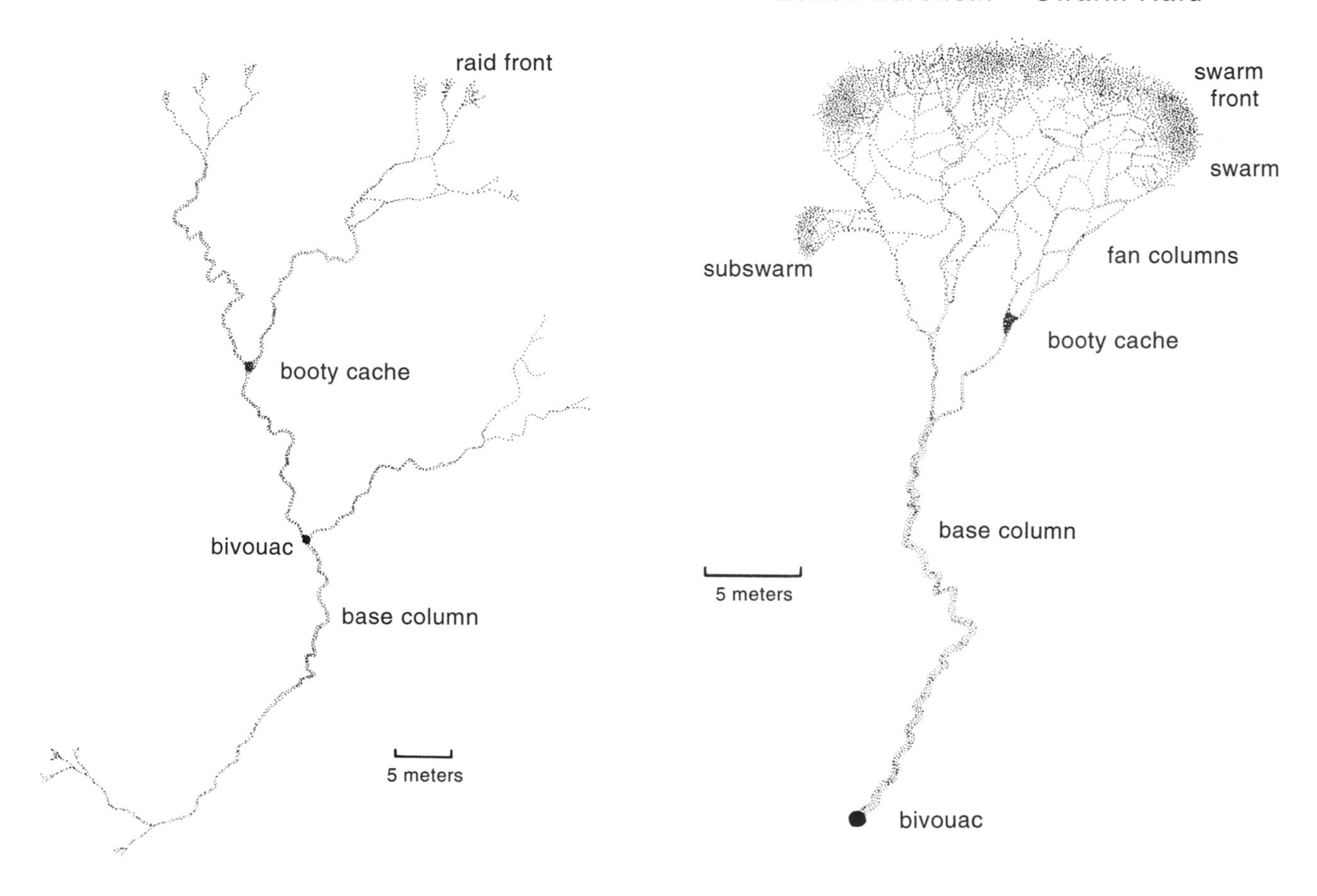

The two basic patterns of raiding employed by army ants. On the left is a column of *Eciton hamatum,* with the advancing front made up of narrow bands of workers. On the right is the unique swarm of *Eciton burchelli,* in which the front is very broad and followed by converging columns in the fan area. (Courtesy of Carl W. Rettenmeyer.)

Thereafter this growing mass holds its initial direction in an appropriate manner through the pressure of ants arriving in rear columns from the direction of the bivouac. The steady advance in a principal direction, usually with not more than 15° deviation to either side, indicates a considerable degree of internal organization, notwithstanding the chaos and confusion that seem to prevail within the advancing mass" (*Report of the Smithsonian Institution for 1955* (1956), pp. 379–406).

Very few animals, large or small, can withstand the approach of the *Eciton* army. Any creature sizable enough to be seized and held in the jaws of the ants must either retreat or die. Colonies of other ants are ploughed under, together with scrambling mobs of spiders, scorpions, beetles, cockroaches, grasshoppers, and other arthropods of great variety. The victims are trapped, stung and torn to pieces, and carried to the rear of the phalanx along the feeder columns to the bivouac, where they are soon eaten. A few arthropods, including ticks and stick insects, are able to protect themselves with repellant secretions that coat their bodies. Termites are mostly safe in their fortress nests of wood and excrement, guarded at the entrances by specialized soldiers with sharp jaws or poison nozzles. But for the most part the swarm raiders fill their role as the unstoppable, superorganismic grim reapers of the tropical forest.

Toward midday the prevailing direction of the workers reverses, and the swarm begins to drain back into the bivouac. The field the ants have covered has been largely depleted of insects and other small animals. As though remembering and sensible of their impact on the environment, the ants on the following morning strike out in a new direction. But if they remain at the same bivouac site for as long as three weeks the food supply will be reduced in all the terrain within easy reach. The colony solves the problem simply by moving at frequent intervals to new bivouac sites a hundred meters or so distant.

Seeing these emigrations in progress, early observers of the tropical environment drew the reasonable conclusion that army-ant colonies change their bivouac sites whenever the surrounding food supply is exhausted. Hunger, it seemed, was the behavioral determinant. In the 1930s, however, Theodore Schneirla discovered that the emigrations are not triggered primarily by empty stomachs but are to some degree

Queens of the army ant *Eciton hamatum*. The individual above, accompanied by a sickle-jawed major worker, is in the nomadic phase; her abdomen is deflated, allowing her to travel easily. The queen below is in the statary phase; her abdomen is swollen with eggs, making it difficult to move from one site to another. (Photographs by Carl Rettenmeyer.)

Slave-raiding in European ants. In this scene, red amazon ants *(Polyergus rufescens)* invade a nest of *Formica fusca* to capture the cocoon-enclosed pupae. Some of the black defenders have picked up pieces of the brood and are attempting to flee. They have little chance of warding off the amazon workers, whose sickle-shaped mandibles easily pierce their bodies. (Painting by John D. Dawson, courtesy of the National Geographic Society.)

Flat-bodied "highwayman beetles," *Amphotis marginata,* on the trails of the European ant *Lasius fuliginosus.* In the left foreground a beetle solicits regurgitation from a food-laden forager. In the upper right, an ant attacks a second *Amphotis,* but has little effect on the turtlelike defense of the beetle. Also present are slender rove beetles (*Pella* species) that hunt and kill the ants. (Painting by John D. Dawson, courtesy of the National Geographic Society.)

The European rove beetle *Lomechusa strumosa* has become fully integrated into the host ant society, in this case *Formica sanguinea.* In this tableau, a *Lomechusa* adult is being fed by a worker while it simultaneously appeases another worker with a calming secretion offered from the tip of its abdomen.

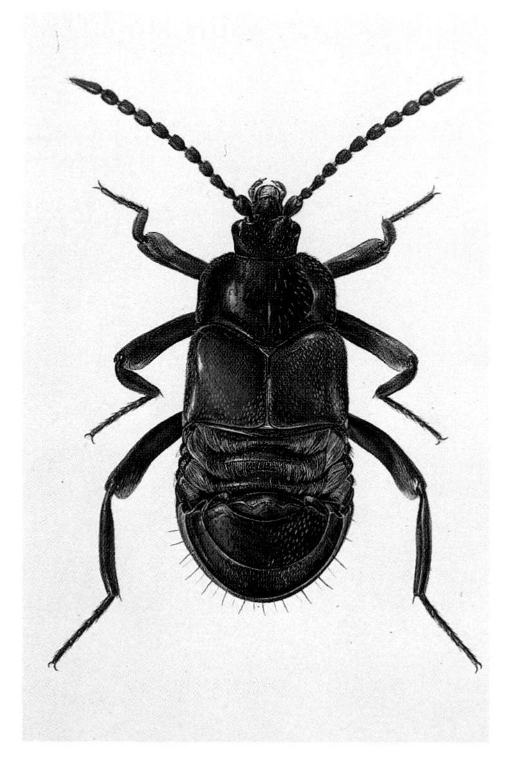
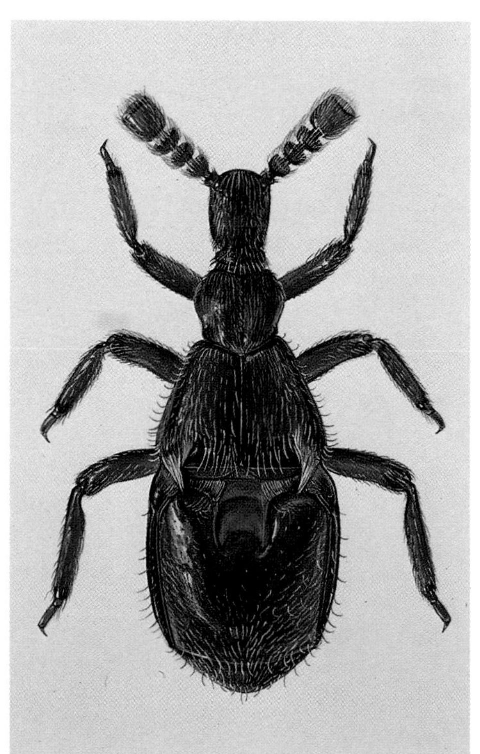

A diversity of myrmecophilous beetles (not shown to scale).
Top left:
Dinara dentata.
Top right:
Lomechusa strumosa.
Lower left:
Claviger testaceus.
Lower right:
Amphotis marginata.
(Paintings by Turid Forsyth.)

Trophobiosis between ants and their homopterous bug guests. The homopterans offer sugary excretions from their anuses in exchange for protection by the ants. *Above:* Australian meat ants *(Iridomyrmex purpureus)* with a nymph of a eurymeline leafhopper. *Below:* African weaver ants *(Oecophylla longinoda)* with scale insects, kept on living twigs inside the ants' arboreal nests.

In addition to lycaenid caterpillars, other butterfly species are associated with weaver ants. This *Homodes* caterpillar, a member of the family Noctuidae, can be seen on trails of *Oecophylla.* The caterpillars are usually not attacked by the ants, but if an ant worker gets too close, the caterpillar exhibits an impressive defensive posture, as shown here with *Oecophylla longinoda* from Africa. The exact nature of the association between the caterpillars and the ants is not known.

Facing page:
In the picture above, taken in Malaysia, weaver ants *(Oecophylla smaragdina)* tend a caterpillar of the lycaenid butterfly *Hypolycaena erylus.* The caterpillar offers a droplet of sugary secretions from a special gland on its back; in return, the ants protect it from enemies. Shown below is an adult *Hypolycaena,* with eyespots and tails that simulate a head; the illusion helps to divert predators from the true head and body while the butterfly escapes. (Photographs by Konrad Fiedler.)

Workers of the nomadic Malaysian ant *Dolichoderus tuberifer* carry their mealybug "cattle" *(Malaicoccus khooi)* to new grazing pastures *(above)*. The ants do not build nests but instead form living shelters with their own massed bodies *(below)*. (Photographs by Martin Dill.)

caused by internal changes that unfold automatically within the colony. The ants move regardless of the richness or poverty of the surrounding food supply. By tracking colonies day after day through the forests of Panama, Schneirla found that the ants alternate between a static period, in which each colony remains at the same bivouac site for as long as two or three weeks, and a nomadic period, when it moves to a new bivouac site at the close of the day, also for a period of two to three weeks. The army-ant colony is driven through this cycle by the internal dynamics of its own reproduction process. The ovaries of the queen develop rapidly after the colony enters the static period, and within a week her abdomen is swollen with about 60,000 eggs, constituting the first of a large batch in production. Then, in a prodigious labor of several days near the midpoint of the static period, the queen lays from 100,000 to 300,000 eggs. By the end of the third and final week of the static period, small, wiggly larvae hatch from the eggs. Several days later the new adult workers of the previous generation shed their pupal skins and emerge en masse from their cocoons. The sudden appearance of tens of thousands of adult workers has a galvanic effect on their older sisters. The general level of activity rises and the size and intensity of the swarm raids grow to a corresponding degree. The colony begins to emigrate to new bivouac sites at the end of each day's raid. Now solidly into its migratory period, it travels the length of a football field each day. This interval of restiveness lasts only as long as the hungry larvae are growing and eating. When they spin cocoons and enter the quiescent pupal period of their development, the colony stops migrating.

Day by day, month by month, the swarm raiders, and with them all the other army ants in the genus *Eciton,* cycle through the same clockwork maneuvers. How can the colony break from such a tight routine in order to reproduce itself? Not easily, given the way the colony makes its living, but reproduction occurs on schedule. Multiplication is foreordained to be a complex and ponderous process. It cannot easily be based upon the mass production and release of winged queens and males, as is the case in most other kinds of ants. Newborn colonies must start with huge numbers of workers supporting each queen. To meet that necessity, a small number of virgin queens are created, which mate without leaving the mother colony. Then one of these queens splits off in the company of an army of workers to form a new colony of her own. The procedure requires a radical realignment of loyalties, so that some workers go with the new queen and others stay with their mother.

Through most of the year the mother queen is the paramount attraction for the workers. By serving as the focal point of the aggregating workers, she literally holds the colony together. The situation changes, however, when the annual sexual brood appears early in the dry season. In the column-raiding army ant *Eciton hamatum* (the species whose reproduction has been most carefully studied) the sexual brood consists of about 1,500 males and 6 queens. The males fly away and enter the bivouacs of other colonies. There they run with the workers and prepare to inseminate the resident virgin queens. Thus is brother-sister incest avoided.

With cross-fertilization now assured, the stage is set for the splitting of the colony. When the next emigration proceeds, one army of workers travels to a new bivouac site with the old mother queen, and another army moves to a second bivouac site in the company of one of the virgin queens. The other virgin queens are left behind, sealed off and prevented from moving by small groups of workers who have chosen to remain behind with her. Deprived of food and defenseless against enemies, the outcasts and their entourages soon die. Within days the successful virgin queen is inseminated by one of the visiting males. The two colonies, mother and daughter, then go their separate ways, never to communicate again.

The 12 known species of the genus *Eciton*, including the swarm-raiding *Eciton burchelli* and column-raiding *Eciton hamatum*, are the furthest extensions of an evolutionary trend that began tens of millions of years ago in the American tropics. Equally interesting to entomologists but far less famous generally are the miniature army ants of the genus *Neivamyrmex*, which range from Argentina to the southern and western United States. In backyards and vacant lots their fierce colonies, hundreds of thousands of workers strong, conduct raids, emigrate from one bivouac site to another, and multiply by fission in the same manner as the *Eciton* marauders. But while they are often literally under foot, people living within their geographic range are almost never aware of

Mating in army ants: a male swarm raider *(Eciton burchelli),* who has clipped wings, copulates with a young queen. (Photograph by Carl Rettenmeyer.)

their existence. At the age of 16, Ed Wilson, already tuned to ant biology, found a colony of *Neivamyrmex nigrescens* behind his family's house near downtown Decatur, Alabama. He watched it for days as it traveled from one site to another, in and out of the weeds along the back fence of the yard, into a neighbor's garden, and then, on one dark rainy day, across the street into still another neighbor's property, where it disappeared. Such a progression in the grass-roots jungle is an exciting spectacle, but it takes patience to tell the legionary forces from foraging columns of the ordinary, more sedentary species, whose nests are firmly established under garden rocks and in the open spaces between clumps of lawn grass. Two years later Ed found other colonies near the campus of the University of Alabama. He used them to conduct one of his first scientific studies, on the strange miniature beetles that ride on the backs of the *Neivamyrmex nigrescens* workers, consuming the oily secretions of the ants for food.

In Africa a second burst of evolution created the fearsome driver ants of the genus *Dorylus,* which we introduced earlier to exemplify the ant colony as a superorganism. A third evolutionary radiation in Africa and Asia created the genus *Aenictus,* miniature army ants superficially similar to *Neivamyrmex.* The behavior and life cycles of these legionary forms are basically similar to those of their American counterparts, yet each of the three evolutionary lines—*Dorylus* and *Aenictus* in the Old World, *Eciton* together with *Neivamyrmex* in the New World—represents an independent evolutionary production. At least that is the opinion of William Gotwald, the American entomologist who has made the most recent study of their anatomy. Gotwald concluded that the similarities are due to evolutionary convergence, not to common ancestry.

Beyond this special group of raiders, other ants have evolved army-ant behavior to one degree or another. The specialization has occurred so frequently, with so many idiosyncratic twists, as to stretch the very meaning of the term "army ant" and require a more formal definition based on what the colonies do rather than on the anatomy of their members. An army ant, to put it succinctly, is any ant belonging to a species whose colonies change their nest sites regularly, and whose workers forage across previously unexplored ground in compact, well-organized groups.

Thus diagnosed in a purely functional sense, army ants of independent ancestry are revealed to occur almost everywhere in the warmer climates of the world. Among the most extraordinary forms are ants in the genus *Leptanilla,* which with several other genera of the Old World make up an entire subfamily of their own, the Leptanillinae. The workers are among the smallest of all ants, so diminutive as to be easily overlooked by the naked eye. Leptanillines are also among the rarest of all species. Neither of us has ever seen a live example, despite years of fieldwork in habitats where the ants certainly occur. Wilson made a special search for them around the Swan River of Australia, where a new species had been discovered 20 years earlier, but without success. William Brown, probably the most widely traveled and productive ant collector of all time, has found only one colony during several years collecting where *Leptanilla* occurs. He came upon it in Malaysia, beneath a piece of rotting wood. The mass of tiny workers shimmered like a rippling membrane on the surface of the wood when first exposed. Brown had to peer at the miniature spectacle for a moment to realize he was looking at ants, and a while longer to realize they were leptanillines.

For a hundred years aficionados of ant evolution speculated that the mysterious leptanillines are army ants. Their anatomy at least vaguely resembles that of the larger, undoubted army ants in the genera *Eciton* and *Dorylus.* But for a long time no one could find and study a colony long enough to test the idea. The breakthrough came in 1987, when a young Japanese myrmecologist, Keiichi Masuko, succeeded in collecting no fewer than 11 complete colonies of *Leptanilla japonica* in the broad-leafed forest at Cape Manazuru, Japan. Each colony, he was able to generalize, contains about a hundred workers and is strictly subterranean—a trait that helps to explain why leptanilline ants are so seldom encountered. To add to their strange nature, the Japanese Leptanillas turn out to be specialized predators on centipedes. This is a hard way to make a living—rather like humans trying to live on tiger steaks. The foragers follow odor trails in a close pack from the nest out to their formidable prey, which are usually many times their size. It is not yet clear, however, whether the centipedes are located by single scouts, which then recruit nestmates, or whether the hunting is undertaken by organized groups in the army-ant manner.

Are the Leptanillas also nomadic? Colonies at home in their earthen nests are certainly restless. They emigrate at the slightest disturbance. The swiftness of their response suggests that they do move at frequent intervals in nature in the manner of army ants. They are also anatomically well adapted for frequent travel. The workers are equipped with special extensions on their mandibles for carrying larvae. The larvae in turn possess a protrusion from the forward part of the body that serves as a handle for the workers to seize, making it easier to haul them from one place to another.

In Japan, Masuko found, the Leptanilla colony undergoes an army-ant cycle of synchronized growth through the warm season. When larvae are present, the colony as a whole is hungry, and the workers hunt centipedes, while apparently moving from one site to another to be near their giant prey. The larvae feast on the centipedes and grow quickly. During

Miniature Asian army ants, *Leptanilla japonica,* use mass attacks to prey on centipedes. When the prey has been subdued, the workers carry the grublike larvae to the centipede so that they can feed on it. In the upper photograph the queen, in the forefront of a typical assemblage, strokes a larva with her antennae. In the lower photograph a queen with an egg-swollen abdomen sits in the midst of a group of grown larvae, holding one in her mandibles. (Photographs by Keiichi Masuko.)

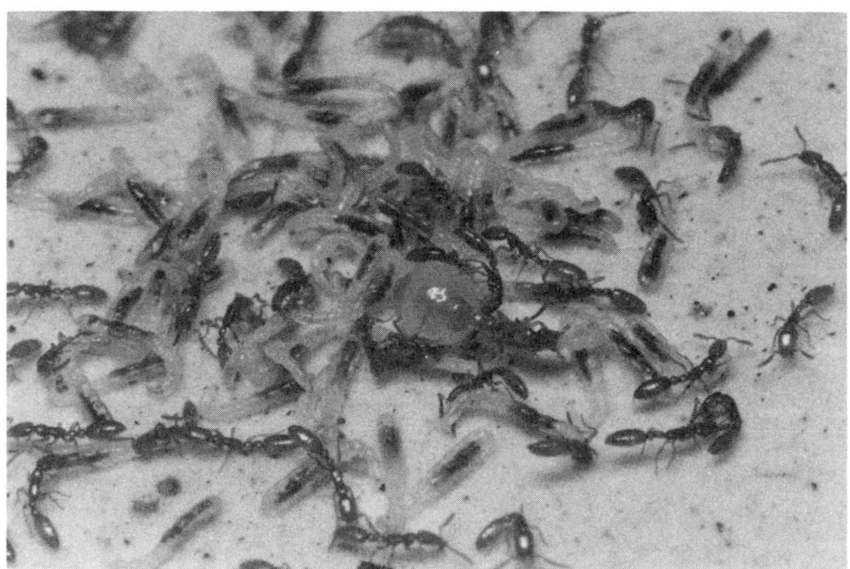

this period the queen's abdomen remains shrunken, and she lays no eggs. While still slender in body form, she is able to run easily with the workers during the colony emigrations. When the larvae reach full size, the queen feeds heavily on larval blood, which is extruded and made available to her through special organs on the abdomens of the larvae. This rich vampire feast causes the queen's ovaries to grow rapidly. Soon her abdomen swells until it resembles an inflated balloon; then within a few days she deposits a large batch of eggs. About the same time the larvae become inactive pupae. With the queen again relatively quiescent, and no larvae to feed, the colony requires much less food. It stops hunting centipedes, and not long afterward settles in for the Japanese winter. In the following spring the eggs hatch into larvae, and the cycle starts anew.

Another bizarre variant of army-ant behavior has recently been discovered in the Asian marauder ant *Pheidologeton diversus* by the American entomologist Mark Moffett. The colonies are huge, containing hundreds of thousands of workers. Unlike the hordes of advanced army ants, they remain in the same nest sites for weeks or months at a time. Yet they conduct swarm raids that are remarkably similar in many respects to those of the African driver ants and the tropical American *Eciton burchelli.*

A *Pheidologeton* raid begins when some ants move away as a group from one of the main odor trails, followed by the rest of the colony. At first the pioneers form a narrow column that grows outward, like water flowing through a pipe, at the rate of up to 20 centimeters a minute. After the column has enlarged to about half a meter to 2 meters in length, some of the ants at its tip start to move laterally from the main direction of the other ants. As a result the swarm slows down, like water spreading in a sheet from the end of the pipe onto the ground. On a few occasions, the expansion strengthens and blossoms into a large, fan-shaped raid. Behind the seething frontal edge of this formation, ants run back and forth through a tapered network of feeder columns. Those returning from the front funnel into a single basal column, which lengthens as the van presses forward into new territory. The swarms each contain tens of thousands of workers. Some travel 6 meters or more from the points of departure. In shape they closely resemble the sallies of the driver ants and American *Eciton* swarm raiders, but they travel across new terrain at a far slower pace.

The Asian *Pheidologeton* marauders, like the more familiar driver and American army ants, are able to conquer exceptionally large and formidable prey, up to and including frogs, by overwhelming them with the sheer force of their numbers. Well-coordinated gangs of workers are able to carry large objects rapidly back to the nest. The ability of the worker force to hunt prey is hugely increased by a complex caste system. The armies contain workers that are the most variable in size of any known ant species: the giant supermajors are 500 times heavier than their smallest nestmates and possess disproportionately massive heads. There is an even gradation of size classes between the two extremes. This diversity allows the swarm of ants to harvest prey of a correspondingly wide range of sizes. Working singly, the smallest of the marauders ferret out springtails and other minute insects that abound everywhere in the soil. Others among the dwarfs join their larger nestmates in savaging termites, centipedes, and other sizable prey. Supermajors move in to deliver the coup de grace with their powerful jaws. These huge ants also serve as the work elephants of the colony, by pushing and lifting sticks and other obstacles out of the way of their onrushing fellow foragers.

During the earlier years of his career, as he traveled widely in the tropics and encountered more army ants, Wilson wondered about the origins of their behavior. How could such an extraordinarily complex social organization originate in evolution? Bit by bit, from his own observations and those of other field biologists such as William Brown, he pieced together the evidences of early evolution in a variety of predatory ants that showed some but not all of the traits of army ants.

A persuasive pattern emerged from this information. The key lay, he found, in the fine details of mass raiding. Earlier writers had pointed out repeatedly that compact armies of ants are superior in the capture of prey to solitary workers. This observation was certainly correct, but it proved to be only part of the story. There is another primary function of group raiding that becomes clear only when the nature of the prey is examined along with how it is captured. Most ants that leave the colony to hunt alone attack prey their own size or smaller. This restriction follows a more general rule of wildlife biology: solitary predators, from frogs and snakes to birds, weasels, and cats, hunt animals their size or smaller. Ants working in groups tend to feed either on big insects or on colonies of ants and other social insects, prey that cannot normally be subdued by a single huntress. They pull the victims down and cut them to pieces by concerted action, just as social groups of lions, wolves, and killer whales hunt the largest mammalian prey.

Many kinds of ants attack large solitary insects and colonies of ants, wasps, and termites in mass raids, yet do not emigrate from one nest site to another at regular intervals like the advanced army ants. These species appear to exemplify the earliest step that led to army-ant behavior. Wilson compared many species that show different degrees of complexity, including the elementary levels. He then was able to reconstruct what he believed to be the origin of the army ants.

In the first step, ants that previously hunted smaller prey in solitary fashion developed the ability to recruit masses of nestmates quickly. The packs specialized on large or heavily armored prey, such as beetle larvae, sowbugs, or the colonies of ants and termites.

Next, the group raids became autonomous. It was no longer necessary for a scout to find the prey first and then recruit gangs of nestmates to subdue it. Now a swarm of workers emerged from the nest simultaneously and hunted as a group from start to finish. This more advanced form of communal raiding allowed colonies to cover a larger area more quickly and to subdue difficult prey before they could escape.

Either at the same time or later, they developed migratory behavior. The efficiency of the group raiders improved, because large insects and colonies are more widely dispersed than other types of prey, and the group-predatory colony must continually shift its hunting area to tap fresh supplies of food. With the addition of regular emigrations, the species became fully functional army ants.

Flexible access to shifting supplies of prey made it possible for the colonies of army ants to evolve to large size. In some species the diet expanded secondarily to include smaller insects and other arthropods, as well as nonsocial insects, and even frogs and a few other small vertebrates. This is the stage reached by the swarm-raiding driver ants of Africa and *Eciton burchelli* of the American tropics, whose colonies sweep virtually all animal life before them. It is reasonable to suppose that these juggernauts of the tropical world, like most of the great achievements of organic evolution, came about through a succession of little steps.

The Strangest Ants

DURING their hundred-million-year history, the ants have pressed to startling extremes of adaptation. Some of the most specialized forms are virtually beyond imagination—they could not have been easily fantasized in advance by the entomologists who were to stumble upon them in the field. What follows is a kind of formicid bestiary of our own making, a set of tales about species we have personally encountered that push the edges of the evolutionary envelope.

Our story begins in 1942, in a vacant lot next to the Wilson family home in Mobile, Alabama. At the edge of the weed-choked property was a fig tree that bore edible fruit late each summer, Mobile being located close to the American subtropics. Under the tree were scattered pieces of lumber, broken glass bottles, and roofing tiles. Around and beneath this rubbish Ed searched for ants. Just turned 13, he had set out to learn all the species he could find. He was startled to find one ant species radically different from anything he had seen before. Medium-sized, slender, dark brown, and very swift, the workers were armed with strange, thin mandibles that amazingly could be opened 180 degrees. When their nest was disturbed, the ants rushed about with their mandibles fully spread. Ed tried to pick them up with his fingers, but they snapped the mandibles shut like miniature bear traps, piercing his skin with sharp teeth, then followed through almost instantly by bending their abdomens forward and delivering a painful sting. So eager were the ants to attack that many of them snapped their mandibles together on empty air, making a clicking sound. The combined one-two punch was shocking. Ed gave up trying to excavate the nest and capture the colony. Later he learned that the species he had found was *Odontomachus insularis,* and that Mobile is at the northern limit of its range. *Odontomachus* is a genus of many species found in the tropics around the world.

Fifty years later, Bert Hölldobler, in the course of research on the predatory ants of the subfamily Ponerinae, began a detailed study of *Odontomachus bauri,* a species closely similar to the one encountered by Wilson. He and his associates Wulfila Gronenberg and Jürgen Tautz at the University of Würzburg became fascinated by

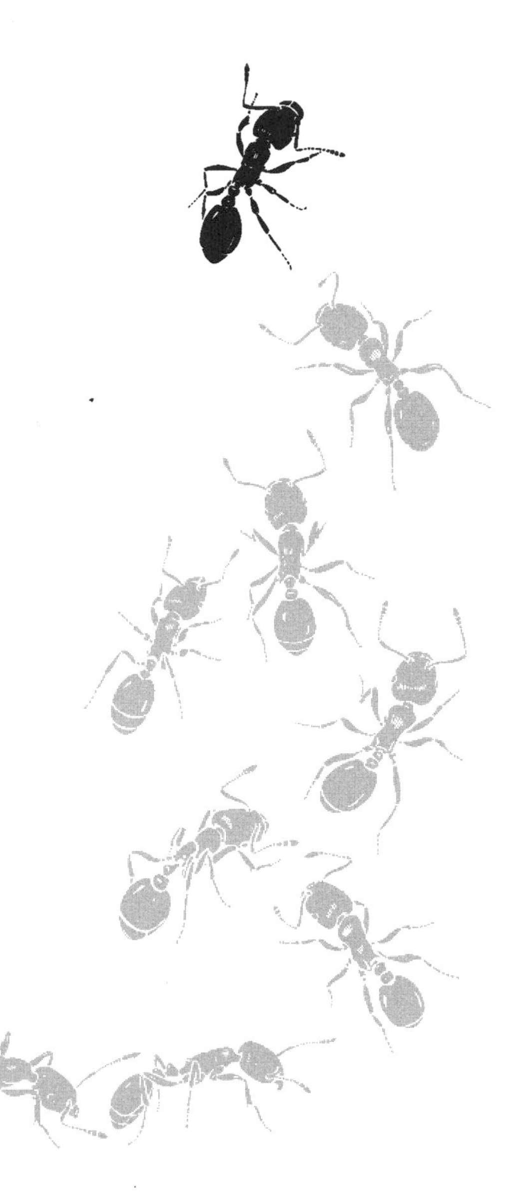

the blinding speed and force of the jaw closure. So hard is the blow that when the tips of the mandibles strike a hard surface the ant flips itself backward through the air. The researchers set out to record the mandibular closure with the aid of ultra-high-speed cinematography, recording at 3,000 frames a second. To their astonishment they found that the movement of the jaws is not merely fast: it is the fastest of any anatomical structure ever recorded in the animal kingdom! The full strike, from the instant at which the fully opened mandibles start to close to the instant they clash together, takes from between a third of a millisecond to a full millisecond—that is, one-three-thousandths to a thousandth of a second. Previously, the fastest recorded movements had been the jump of a springtail at 4 milliseconds, the escape response of a cockroach (40 milliseconds), the foreleg strike of a preying mantis (42 milliseconds), the tongue "shoot" of a rove beetle to catch prey (1–3 milliseconds), and the leap of a flea (0.7–1.2 milliseconds). The *Odontomachus* mandible is only 1.8 millimeters long, but its spiked tip moves at a velocity of 8.5 meters a second. If the ant were human, its response would be the equivalent of swinging the fist at about 3 kilometers a second—faster than a rifle bullet.

The *Odontomachus* trap-jaw workers can catch any known living creature, provided they can get it within range of their mandibles. They hunt with their jaws open and locked into position, ready for the pull of the massive adductor muscles. A long sensitive hair projects forward from the base of each mandible. During the hunt the *Odontomachus* worker sweeps her antennae back and forth in front of the head. When the smell organs on her antennal surface identify either a prey or an enemy, the ant jerks her head forward, causing the tips of the hairs to touch the target. Inside the mandibles are huge nerve cells that respond to pressure on the hairs. Their axons, the elongated cell stems, are the largest ever recorded in either insects or vertebrates. Their size, Hölldobler's co-workers found, allows them to conduct impulses at extremely high velocity. The reflex arc, running from the receptor cells in the jaw to the brain and back out to the motor cells of the mandibular muscles, takes only 8 milliseconds, the shortest duration recorded in any animal to date. When the electric discharge completes the arc, so that the impulse

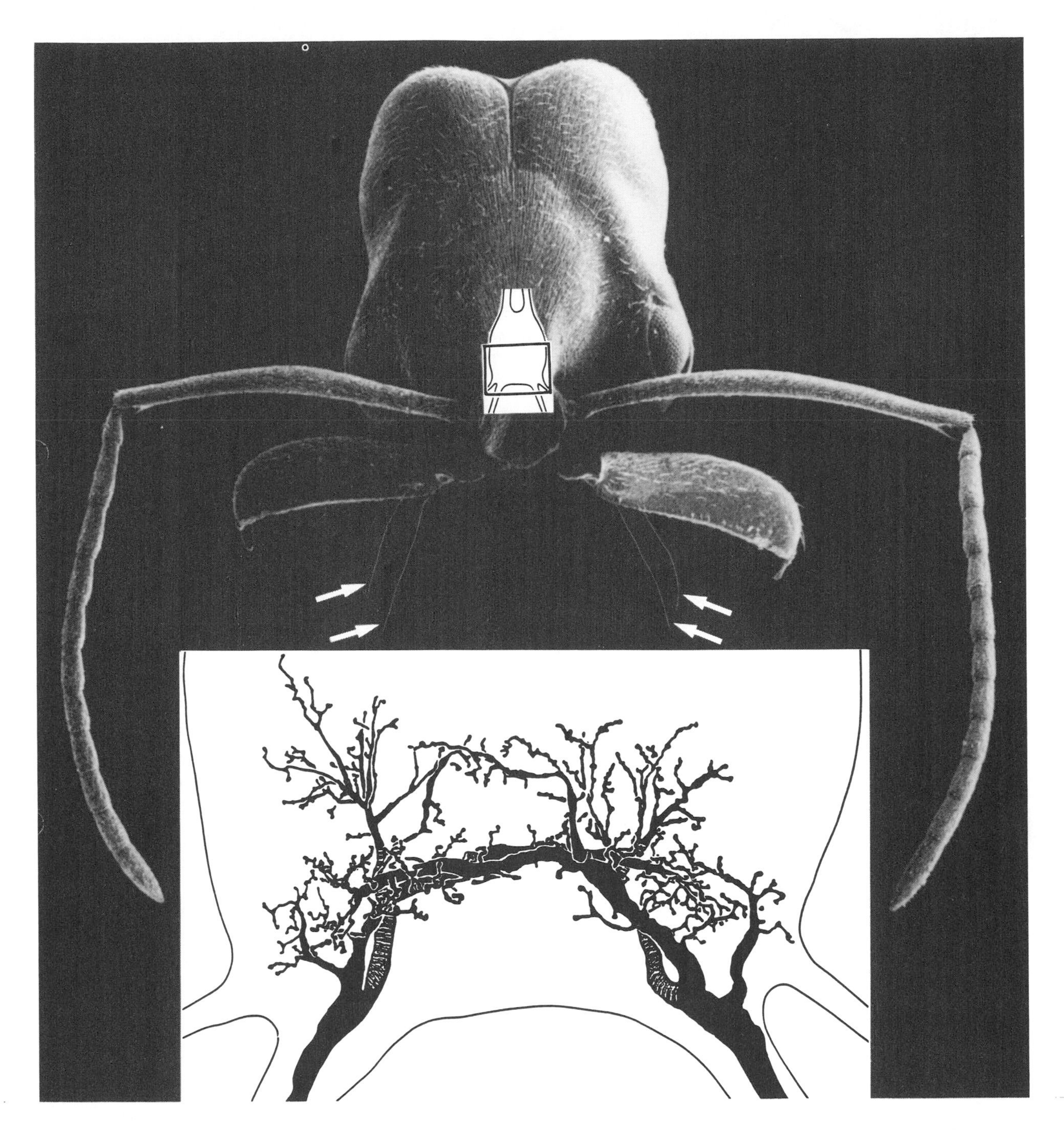

The powerful and swift trap jaws of *Odontomachus.* Here the mandibles of a worker have been opened fully; the arrows point to the sensitive trigger hairs that project forward. The inset reveals the portion of the brain entered by the gigantic nerve cells from these trigger hairs. In the lower drawing this portion of the brain is enlarged, and the nerves are shown in black. (Drawing by Wulfila Gronenberg.)

reaches the muscles, the mandibles close within a millisecond, completing the full behavioral response.

The *Odontomachus* mandibles are filled mostly by giant sensory cells surrounded by air space; the resulting lightness enhances their startling velocity. When they snap together the jaws stun smaller creatures or at least pierce them with the terminal teeth, holding them still while the ant bends its abdomen forward to insert its sting. The mandibular strike is powerful enough to cut some soft-bodied insects in half.

The superfast mandible strike of *Odontomachus* also serves a second, wholly different function. The workers use it as a transport device when attacking intruders. By pointing their heads downward toward a hard surface and closing the mandibles, the ants are able to catapult themselves up into the air—and onto nearby enemies. When Bert Hölldobler touched a nest of a large *Odontomachus* species in a tree at La Selva, Costa Rica, as many as 20 workers snapped their mandibles and sailed about 40 centimeters through the air onto his body. As soon as they landed they began stinging him. Bert stepped back involuntarily. He understood at once how the colonies of this species protect their otherwise vulnerable homes, the walls of which are little more than a thatching of dry vegetable materials.

Other ants with heads like snares are abundant in tropical and warm temperate regions around the world. The *Odontomachus*-like armamentarium has originated many times independently in the course of evolution. As a college student in the late 1940s, Wilson turned his attention to one of these groups, the dacetines, small ants known to hunt springtails. Many species live in Alabama, including members of the genera *Strumigenys, Smithistruma,* and *Trichoscapa,* which until that time had remained almost entirely unstudied. Wilson set out to find all the dacetines he could, searching through woodlands and fields in the central and southern parts of the state. He housed the colonies, which were typically composed of a single queen and a few dozen workers, in artificial nests made from blocks of plaster of Paris. The construction was modified from a design introduced a half century earlier by the French entomologist Charles Janet. To observe his dacetines as closely as possible Wilson carved holes in half of the upper surface to create little chambers and connecting galleries similar to those excavated by the ants

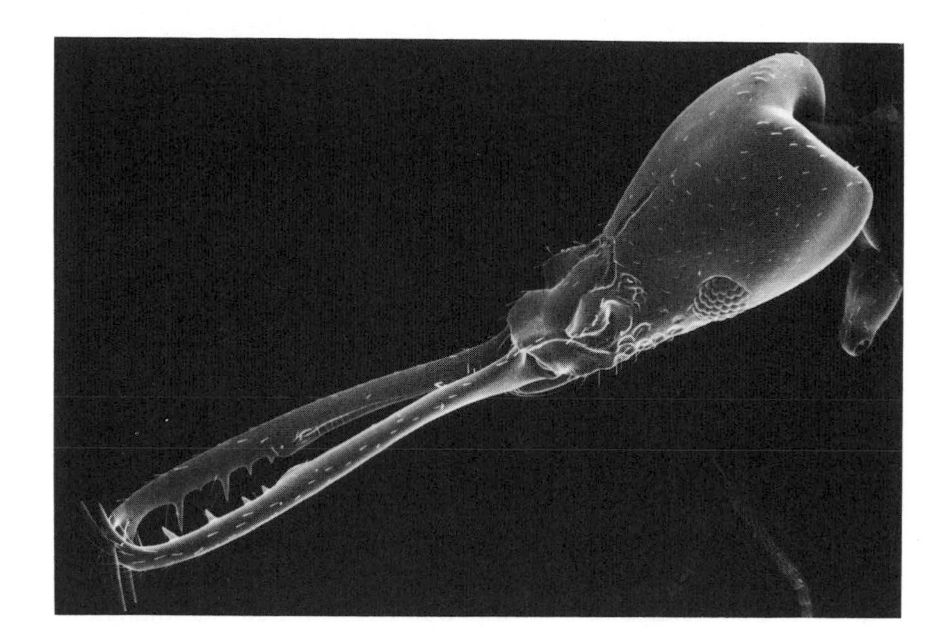

A trap-jawed worker of a Central American species of *Acanthognathus;* the extremely long and slender mandibles are used to capture springtails and other small, swift insects.

themselves. Into the other half he dug a much larger chamber to serve as the foraging arena of the ants. Then he covered the entire surface with a plate of glass to create a transparent roof. On the floor of the arena he scattered bits of soil and decayed wood to simulate the natural floor of the forest. Into this space, finally, Wilson placed live springtails, mites, spiders, beetles, centipedes, and other small arthropods collected from habitats where the dacetines lived, in order to see which would be hunted by the dacetines—and in what manner. The entire block of plaster was small enough, about the size of two doubled fists, to be fitted onto the stage of a dissecting microscope. So with minimal effort Wilson was able to watch both the dacetine colony in the brood chambers and the workers hunting in the foraging arena, virtually all at the same time.

There are two basic kinds of trap-jawed ants among the dacetine species. One has extremely long, slender mandibles, which the ants open 180 degrees or more, as in *Odontomachus,* and then close with a convulsive snap, impaling the prey on the sharp terminal teeth. The ants move about a great deal while hunting, and stalk insects only for relatively brief periods of time once they have located them. The second group has shorter jaws, which can be opened only about 60 degrees. These short-mandibled dacetines, Wilson discovered, are masters of stealth. As soon as a huntress becomes aware of an insect nearby, she freezes in a crouching position and holds this stance briefly. Then if at an angle to the prey, she slowly turns to face it. She next begins to creep forward, in a movement so slow that it can be detected only by persistent and careful watching and by noting the position of her head relative to bits of soil. Several minutes may pass before the ant comes into striking position. If the prey moves in any way during the stalk, the dacetine freezes again and waits a while before resuming forward movement. Finally she comes within range, whereupon she touches the prey very gently with the tips of the long sensitive hairs that project forward from her head, and closes her mandibles with an explosive snap.

The dacetines studied by Wilson in his miniature terraria displayed a general liking for small soft-bodied arthropods, including symphylans, which resemble centipedes, and diplurans, which look like miniature silverfish. But most of all they relish springtails, tiny wingless insects that possess a forked, taillike appendage beneath the body (the furcula) allowing them to bound instantly and far away at the slightest hint of danger. The release and downward swing of the furcula is one of the fastest movements known in the animal kingdom, exceeded only by the mandibular snap of the *Odontomachus.* The small dacetine ants, using stealth and a convulsive snap of the trap jaws, are among the few animals able to capture springtails consistently.

In later studies Keiichi Masuko, known for solving the riddle of the leptanilline army ants, used his extraordinary powers of observation to add a new twist to the dacetine story. The little workers, he found, smear soil and other detritus on their own bodies, evidently as an odor camouflage to allow a closer approach to the prey. In still another extension, Alain Dejean of France found that the workers exude an odor attractive to springtails, holding them in place longer as the ants make their approach.

Over the years Wilson and William Brown, in the course of hunting dacetine ants during expeditions to various parts of the tropics, pieced together the likely evolution of the miniature stealth hunters. About 250

Two trap-jawed ants: a species of *Myrmoteras* from southeastern Asia *(above)* and *Daceton armigerum* of South America *(below).*

dacetine species are known worldwide, composing 24 genera, and they vary enormously among themselves in size, anatomy, and behavior. Their history evidently proceeded as follows. The more primitive forms, like the living species of *Daceton* in South America and *Orectognathus* in Australia, were large ants that foraged over the ground and onto low vegetation. They used trap jaws to catch a wide variety of small to moderate-sized prey, such as flies, wasps, and grasshoppers. In some lines originating from these ancestral forms, the workers drastically reduced their size and began to hunt minute, soft-bodied insects and other arthropods living in the soil. Some of the extreme species came to restrict themselves entirely to the capture of springtails. At the same time the social structure was altered to accommodate this change to a reduced, secretive lifestyle. The colonies grew smaller, the workers became uniform in size (in contrast to the maintenance of major and minor workers in the larger dacetines), and the ants abandoned the use of odor trails to recruit nestmates to prey.

This sketch of dacetine history, mostly completed by 1959, was one of the first attempts to reconstruct the evolution of social organization in a group of animals as shaped by changes in food habits and other dimensions of ecology.

The jaws of an ant are the functional equivalent of the hands of a human being. They are used to pick up and manipulate soil particles, food items, and nestmates. They serve as weapons for the defeat of enemies and the capture of prey. The size and shape of the jaws therefore provide clues to the kinds of lives ants live and the nature of the food the workers gather. And of all the ants in the world, those with the strangest mandibles are not *Odontomachus* and the dacetines but the species of the ponerine genus *Thaumatomyrmex*. The head capsule of the worker is short, almost globular, and set on each side by large, convex eyes. The huge mandibles project forward like a basket, the struts formed of long thin teeth resembling the tines of a pitchfork. When the mandibles are closed tightly against the mouth in repose, the tips of the extremely long terminal teeth extend past the rear border of the head like a pair of horns. The name *Thaumatomyrmex* means, appropriately enough, "marvelous ant."

Marvelous indeed—and just how are those spectacular jaws used? Are they trap jaws, or do they fill some other, wholly unsuspected role? For years myrmecologists speculated on the natural history of *Thaumatomyrmex*—where it nests and the creatures it hunts. Unfortunately, members of the genus are among the rarest ants in the world. Although the several known species are distributed variously from southern Mexico to Brazil (with one found only on Cuba), no more than a hundred specimens exist in all the museums of the world. Just to find a single living worker is considered a major accomplishment. Until recently no live colony had ever been studied in the laboratory.

Wilson had managed to collect just two workers in his entire life, one in Cuba and the other in Mexico. It was his burning ambition for many years to locate a colony and solve the mystery of the stupendous jaws. In 1987 he devoted a week to this task at the La Selva Field Station and Biological Station of the Organization for Tropical Studies, in northeastern Costa Rica, a locality where several specimens had been recently collected. During that time he did nothing but walk along trails and across the undisturbed forest floor, head bent, desultorily kicking at leaves and fallen tree branches, in search of the distinctive black, shiny form of the basket-headed workers. He found not a single one. In frustration, he next published an article in *Notes from Underground,* the myrmecological newsletter. The message in essence: "Will someone please find out what *Thaumatomyrmex* eats, and put my mind at rest?"

Within a year three young Brazilian scientists, C. Roberto F. ("Beto") Brandão, J. L. M. Diniz, and E. M. Tomotake, had the answer. They came across two workers, in separate locations in Brazil, carrying dead polyxenid millipedes. They also located a fragment of a colony and kept it under observation in the laboratory, during which time the workers accepted polyxenid millipedes while ignoring other kinds of prey offered them. Millipedes have two legs on every segment of their body, and are sometimes called thousand-legs. Most are elongate, cylindrical creatures with hard, calcareous exoskeletons. Polyxenid millipedes are very different in general appearance, however. Relatively short, soft-bodied, and covered by long, densely packed bristles, they are the porcupines of the millipede world.

The *Thaumatomyrmex* are porcupine hunters. Their extraordinary

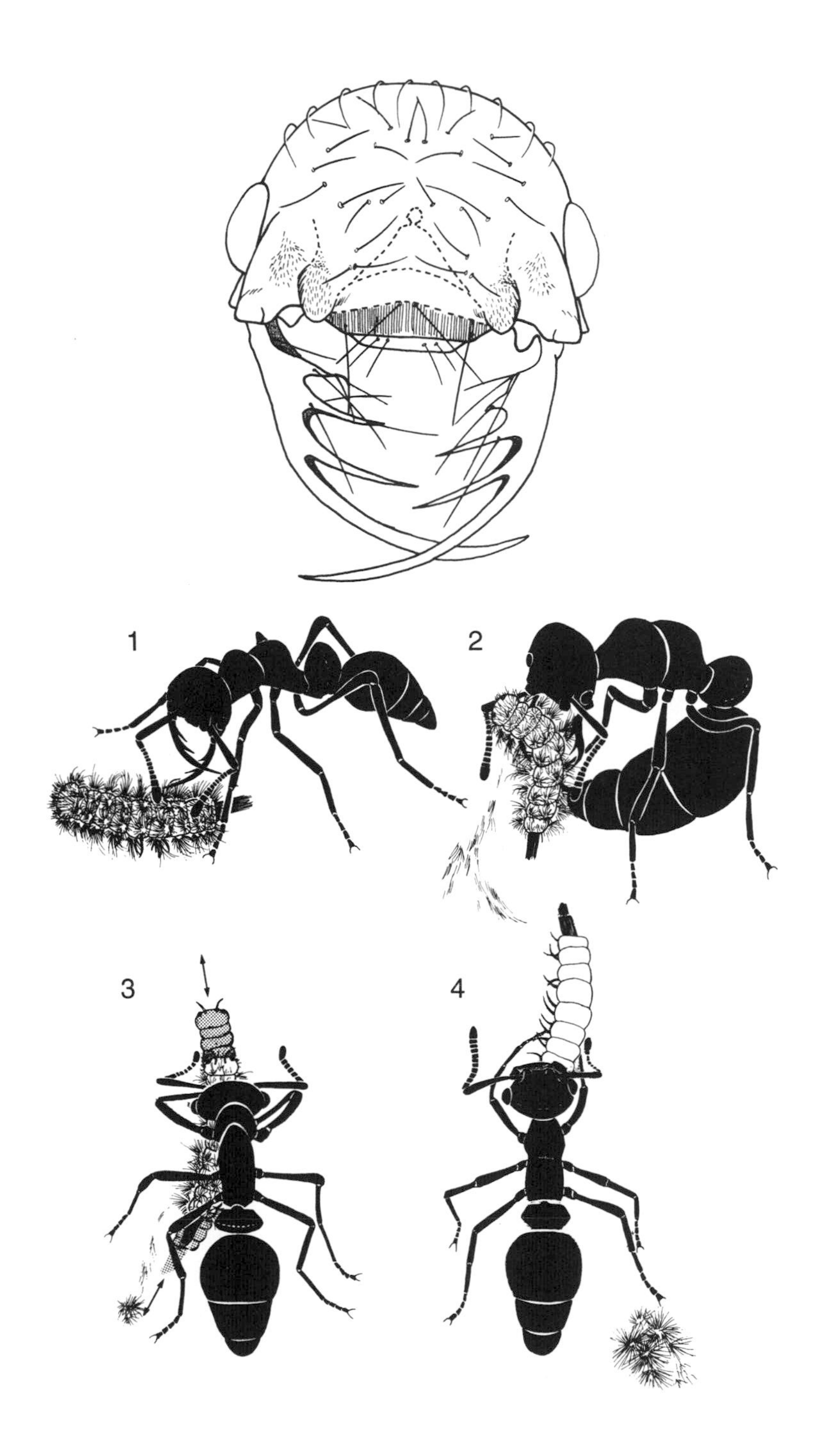

The tropical American species of *Thaumatomyrmex* are among the rarest ants in the world. They also have the most bizarre mandibles, with which they capture porcupine-like polyxenid millipedes. In the sequence below, a worker strips the bristles from a polyxenid before dismembering and eating it. (From an article by C. R. F. Brandão, J. L. M. Diniz, and E. M. Tomotake.)

mandibles are nicely adapted to overcome the defense of the polyxenids. Brandão and his collaborators learned that on encountering one of the millipedes, the ant drives the spikelike teeth of her mandibles past its bristles and into the body, and then carries her prey home. Inside the nest she uses coarse hairs on the pads of her forefeet to strip off the bristles of the millipede, like a cook plucking a chicken for the pot. Then the ant consumes the millipede, starting at the head and proceeding back to the tail. Occasionally it shares some of the remains with adult nestmates and the larvae. On learning of this startling discovery, Wilson was happy to know the secret of *Thaumatomyrmex* at last, disappointed not to have come close to discovering or even guessing the answer himself, and at another level a bit sad to know that the ant netherworld had one less challenge to offer.

Another mystery recently solved was the natural history of big, dark ants in the genus *Basiceros*. The workers' heads are elongate, their integument thick and coarsely sculptured, and their bodies cloaked with a mixture of bizarre club-shaped and feathery hairs. Like the pitchfork-jawed *Thaumatomyrmex*, the species of *Basiceros* are widely distributed in the forests of Central and South America, yet until recently few had been seen alive. Virtually nothing was known about their natural history.

The rarity of *Basiceros* turns out to be an illusion, because the ant itself is a master of illusion. In 1985, while collecting in the biological reserve at La Selva, we learned how to locate colonies of the local species with relative ease. We discovered that the ant, *Basiceros manni*, is actually quite common. The trick in hunting it is to look for the white larvae and pupae, which stand out from the dark rotting wood in which the ants build their nests. The workers and queens themselves are extremely difficult to locate unless one knows exactly where to look and then stares at that spot closely. The ants are superbly camouflaged to the human eye, and presumably also to the eye of visually searching predators such as birds and lizards. The ants are easily lost sight of as they walk over the ground, and they become virtually invisible when they come to a halt. The effect is achieved in part by the extreme slowness of the *Basiceros manni* workers. They are among the most sluggish ants we have encountered during years of field experience around the world. They are slow-motion huntresses, creeping about in search of insects, which they stalk carefully and seize with a sudden snap of their jaws. Inside the nest, the entire worker force often stands perfectly still for minutes at a time, even holding their antennae rigidly in place. For the observer used to the eternal bustle of most ant colonies, the effect is eerie. When moving workers are disturbed by being uncovered or touched by a pair of forceps, they freeze into immobility for up to several minutes, in contrast to most other kinds of ants, which dash frantically away.

Not only are the *Basiceros* phlegmatic in the extreme, they are also the dirtiest ants in the world. Most ants are scrupulously clean. They stop frequently to lick their legs and antennae and wipe their bodies with combs on their legs and brushes of hairs on their feet. In some species more than half the behavioral acts of the ants are devoted to cleaning their own bodies, and a large portion of the remainder involves washing nestmates. The *Basiceros* devote only 1 to 3 percent of their repertoire to personal grooming. The bodies of older workers are encrusted with dirt. The phenomenon is not the result of neglect and poor hygiene, but is a quality sought by the ants. It is part of the camouflage technique of the species. By the time the workers are old enough to forage outside the nest, they blend in almost perfectly with the soil and rotting litter they walk on.

The camouflage of the *Basiceros* is enhanced by anatomical design. The collection of fine particles is accomplished by two layers of hairs on the upper surface of the body and legs. Long hairs with splintered ends, shaped like bottle brushes, scrape off and capture minute soil particles. Beneath them, like bushes in a forest undergrowth, feather-shaped hairs anchor the particles close to the body surface.

We successfully cultured *Basiceros* colonies in artificial nests back at Harvard by feeding them vestigial-winged fruit flies, which the ants hunted in their usual slow motion. The workers had no natural dark soil to add to their exoskeletons, but they did have the fine dust of the plaster of Paris walls and floor from which we had built the laboratory nests. And so over time the older workers began to turn white—they became ghost ants, camouflaged in an environment in which no *Basiceros* had previously lived.

The exact opposites of the stealthy dacetines and *Basiceros* are ants that flaunt bright colors in sunlight. They subscribe to a basic rule of natural history that holds both on the land and in the sea: if an animal is beautifully colored and acts with relative indifference to your presence, it is probably poisonous or well armored with jaws or spines. On the floors of Central and South American rain forests, the bodies of poison-arrow frogs are dazzling spots of color in various combinations of red, black, and blue. The frogs make only a half-hearted attempt to hop away when approached, and they sometimes sit still if you try to pick them up. Don't. The mucus of a single individual is toxic enough when ingested to kill a human being. Amerindian hunters, in order to immobilize monkeys and other larger animals, add a trace of the material to the tips of their arrows and blow darts.

In Australia, red and black bulldog ants, a centimeter or more in length and packing a sting as powerful as that of a wasp, can be spotted from a distance of 10 meters. Around their nests they are fearless and belligerent, and they have excellent vision. The workers of some species literally bound toward human intruders, taking sizable forward leaps into the air.

Some of the most colorful and insouciant ants in the world occur in Cuba. They are members of the genus *Leptothorax*, until recently placed in a genus of their own, *Macromischa*, because of their special anatomical properties. There are dozens of species on the great island, almost all of which are found nowhere else. Jewels of Antillean natural history, they come in many sizes, shapes, and colors, including yellow, red, and black. But the most arresting of all are slender species that shine metallic blue and green in the sunlight. The workers forage in open spaces, often in columns, on limestone walls and low woody vegetation.

When Wilson was 10 years old, he was enchanted by the following passage in a *National Geographic* article by William Mann: "I remember one Christmas Day at the Mina Carlota, in the Sierra de Trinidad of Cuba. When I attempted to turn over a large rock to see what was living underneath, the rock split in the middle, and there, in the very center, was a half teaspoonful of brilliant green metallic ants glistening in the sunshine. They proved to be an unknown species."

Imagine! Prospecting in a faraway place for new species of ants that resemble living emeralds. Mann named the species *Macromischa wheeleri,* in honor of his Ph.D. adviser at Harvard, William Morton Wheeler. The image was still vivid in Wilson's mind in 1953 when, as a Ph.D. student at Harvard, he arrived at the same locality, Mina Carlota, to collect ants. As he climbed a steep, forested hillside, he turned over one soft limestone rock after another in search of ants, as had Mann. Some cracked, some crumbled, and most stayed intact. For a while, no green ants appeared. Then one rock broke in half, exposing a teaspoonful of the metallescent workers of *Leptothorax wheeleri.* There was a special satisfaction in repeating Mann's scientific discovery in exact detail four decades later. It was a reassurance of the continuity of the natural world, and of the human mind.

Continuing deeper into the Sierra de Trinidad, Wilson encountered another *Leptothorax* species whose workers glistened with golden reflections in the sunlight. The color resembled the scintillations of tortoise beetles found in many parts of the world. The hue (as well as the metallic blues and greens of other species) is almost certainly produced by microscopic ridges on the body that refract strong light. But why should such an unusual effect be evolved in the first place? It is a fair guess that the ants are also poisonous and thus use the color to warn off predators, perhaps the anole lizards that also abound in the same habitats. A few other ants in the world have turned golden. Some species of *Polyrhachis* in Australia and Africa have evolved sheets of golden hairs on their abdomens, which may serve to advertise the sharp spines they carry on their thoraxes and waists.

Let us close our bestiary with absolutely the rarest, or at least the most elusive, of all the ants we have ever known. In 1985 Hölldobler was searching along the edge of second-growth woodland at La Selva, our favorite tropical study site. He poked at a curious small cluster of dried, thatchlike vegetation about chest high in the foliage of a small tree. Out poured more than a hundred workers of a new species belonging to the ant genus *Pheidole.* The ants ran in erratic looping patterns to form a spreading pattern away from the nest. There was nothing unusual about this response—ants usually rush out to defend their homes—except that in this case the workers looked astonishingly like termites of the genus *Nasutitermes.* These termites are abundant in the trees of La Selva and elsewhere throughout the New World tropics. Their huge spherical nests, built from hardened feces, contain tens of thousands of workers. The soldier caste, called nasutes, possess long, noselike projections on the head from which they squirt streams of sticky, noxious fluid. The nasutes swarm out in large numbers whenever the nest walls are ruptured. Few enemies the size of frogs or smaller can withstand their attacks.

The *Pheidole* discovered by Hölldobler resembled the nasutes only superficially, but the deception was convincing. At first he thought they were actually termites. The movements of the charging workers were nearly identical to those of *Nasutitermes.* Even more persuasively, the coloration of the *Pheidole* soldiers is unique for ants of this genus but close to that of the termite soldiers. If our interpretation is correct, the ant, which we later named *Pheidole nasutoides,* is the first known case of an ant mimicking a termite. In this role, it presumably bluffs predators that have learned to avoid the heavily armed nasutes.

For the remainder of our visit to La Selva that year we searched hard for more colonies of *Pheidole nasutoides,* in order to make a close study of its natural history and to test the mimicry hypothesis. But we never found another. In later trips, sometimes separately and sometimes together, we continued the search, but still had no luck. We are puzzled by this failure, and eager to learn more about *Pheidole nasutoides.* It is possible that the ant is simply very rare, limited to extremely sparse populations like those of the spear-toothed *Thaumatomyrmex.* Or it may be normally a dweller of the high canopy, a zone we and others have yet to explore. Perhaps the nest had fallen from a branch higher up. Eventually someone will learn the answer, and the puzzle will be solved. There is no need to fear that the world of ants will then grow one bit less interesting. By that time other strange phenomena will certainly have come to light, leading new generations to adventures in the field.

How Ants Control Their Environment

ANT COLONIES control and change the environment to their liking by means of mass action and division of labor among the workers. Temperature regulation is a prime example of this social power, one vital to the success of the ants. For some reason still unknown, these insects require an unusual amount of heat. With the exception of the primitive Australian *Nothomyrmecia macrops* and a very few other cold-temperate species, ants function poorly below 20°C (68°F) and not at all below 10°C (50°F). Their diversity declines steeply from the tropics to the north temperate zones. Colonies of any kind are scarce in the shaded portions of old-growth northern coniferous forests, and only a very few cold-adapted species live on the tundra. No native species of any kind exists in Iceland, Greenland, or the Falkland Islands. Ants are also largely absent from the slopes of heavily forested mountains in the tropics above 2,500 meters (8,200 feet). In contrast, a legion of species swarm in the hottest and driest places, from the Mojave and Sahara Deserts to the dead heart of Australia.

In cool habitats ants seek heat for the rearing of their larvae. This, in simplest terms, is why colonies are concentrated so heavily beneath rocks in the cold temperate zone, and why the best way to find entire colonies with the queen near the surface is to turn over rocks, preferably in the spring when the ground is first warming up. Rocks have excellent thermoregulatory properties, especially those that are flat and set shallowly in the soil, with a large fraction of their surface exposed to the sun. When dry they have a low specific heat, meaning that only a small amount of solar energy is needed to raise their temperature. Hence during the spring, when ant colonies most need to move into action quickly, the sun warms the rocks and underlying soil more rapidly than it does the surrounding soil. The difference allows the workers to forage, the queen to lay eggs, and the larvae to develop sooner than rivals confined to bare soil. The same thermoregulatory principle holds for the spaces beneath the bark of decaying stumps and logs. In spring the queen, workers, and brood crowd together in such cavities, while retreating through passageways to the cool interior of the wood only when the outer chambers become overheated.

Ant species in tropical forests, enjoying a sufficient warmth almost all the time, display a very different nesting preference. Most inhabit small pieces of rotting wood on the ground. A small number nest in bushes and trees or in rotting logs, and still fewer live entirely in the soil. Where rocks occur on the ground, they are seldom chosen by ants for cover.

The complete adaptation to ground life of ants gives them a special opportunity to regulate their surrounding temperature on an hourly basis. Their nests are typically excavated from beneath rocks or the bare ground surface vertically into the soil, or else from spaces beneath the bark of rotting wood into the heartwood and around the heartwood surface to encompass portions of the wood facing the soil. This geometry allows workers to move the eggs, larvae, and pupae quickly within the nest to reach the chambers best suited for growth. Colonies of most species manage to keep all stages of brood in the warmest chambers within 25°-35°C whenever these temperatures are available.

The earthen nests also permit the ants to avoid overheating in the hottest environments. Even desert specialists die if forced to stay above ground in the summer sun for more than two or three hours. Surface temperatures above 50°C, which are reached in some deserts, cause death within minutes or even seconds. The ants nevertheless manage to flourish by constructing nests deep within the soil, where temperatures stay close to a comfortable (for ants) 30°C even on the hottest days.

The most sophisticated climatic regulation is achieved by ants that build mounds. These structures are far more than piles of earth excavated to create large underground dwellings. They are intricate in design, symmetric in shape, rich in organic materials, perforated with dense systems of interconnected galleries and chambers, and often thatched with fragments of leaves and stems or sprinkled with pebbles and bits of charcoal. True mounds are cities above the ground, filled with ants and their brood. They are found most commonly in habitats subject to extremes of temperature and humidity, such as bogs, stream banks, coniferous woodlands, and deserts.

The mounds best understood through research to date are the large structures created by ants of the genus *Formica* in cold temperate zone climates. The massive constructions of the red and black wood ants, including *Formica polyctena* and closely related species, are familiar sights in the forests of northern Europe. Rising as much as 1.5 meters (5 feet) above ground level, the mounds are designed to raise the temperature of the ants inside, which are then able to forage earlier in the spring and rear new broods more quickly. The outer crustlike layer reduces loss of heat and moisture, while the enlarged area of the surface exposes the nest to more sunlight. The mounds of some *Formica* species also have longer southern slopes, which further increase the amount of solar energy collected. These slopes are so consistently oriented that for centuries the nests have been used as crude compasses by natives of the Alps. Additional heat comes from the decay of plant materials gathered within the mound and from metabolism of the tens of thousands of ants working together in crowded quarters.

Some ants, such as the *Pogonomyrmex* harvester ants of the American deserts and grasslands, decorate the surface of their mounds variously with small pebbles, fragments of dead leaves and other vegetation, and pieces of charcoal. These dry materials heat rapidly in the sun and serve as solar energy traps. On the high plains of Afghanistan, colonies of *Cataglyphis* sprinkle their mounds with small stones. The habit may be the basis of the legend of ant gold mining reported by Herodotus and Pliny. Herodotus placed the Afghan mining ants near the town of Caspatyros in the country of Pactyike, which has been identified as either modern-day Kabul or nearby Peshawar. It is well known that gold is found in rock and alluvial soil in this part of Afghanistan, and nuggets might occasionally have been brought to the surface by ants along with the pebbles used in temperature regulation. In a somewhat similar fashion, *Pogonomyrmex* harvester ants in the western United States regularly add fossil bones of small mammals to the outer decoration zones of their nest surfaces. Paleontologists routinely inspect the mounds early in their expeditions to see if there are any skeletons still buried in the vicinity.

The greatest peril of the physical environment faced by ants is not excessive heat or cold or drowning (most can live under water for hours or even days), but drought. Colonies of most species need an ambient

A foraging *Pachycondyla villosa* worker brings a water droplet into the nest, where it will be shared with nestmates and daubed onto the walls and floor of the nest to humidify the interior.

humidity higher than that of ordinary outside air, and they face death within hours if exposed to very dry air. Ants therefore employ a diversity of techniques, some approaching the bizarre, to raise and regulate humidity in the nest chambers. Mounds, for example, appear to be constructed to keep not just the temperature but also the moisture of the air and soil within tolerable limits. The thick crust and thatching reduce evaporation, and in addition nurse workers move the immature forms up and down through the vertical passageways to reach optimum humidity. They place the delicate eggs and larvae in moister rooms and the pupae in the drier ones, the latter usually closer to the surface.

A radically different form of humidity control is practiced by *Pachycondyla villosa,* a giant ponerine hunting ant found from Mexico to Argentina. During the dry season colonies living in arid habitats are in constant danger of desiccation. Gangs of workers make repeated trips to collect dew from nearby vegetation or water from any other source they can find. They gather the droplets between their widely opened mandibles and carry them back to the nest, where they pause and allow thirsty nestmates to drink some of the excess. The remainder of the water is then fed to larvae, daubed onto cocoons, and placed directly onto the ground. Using this bucket brigade the *Pachycondyla* foragers keep the interior of the nest much moister than the surrounding soil.

A strange variation on water collecting is used by the Asiatic hunting ant *Diacamma rugosum.* In the dry scrub woodland of India workers decorate the entrances of their nests with highly absorbent objects such as bird feathers and dead ants. In the early morning hours the dew forming on this material is gathered by the *Diacamma* workers. During the dry season the droplets appear to be the only source of water for the ants.

Still another and equally strange form of humidity control is "wallpapering" by *Prionopelta amabilis,* a tiny, primitive ponerine found in Central American rain forests. The colony typically constructs nests in logs and other fragments of rotting wood on the forest floor, materials that are saturated with water a large part of the year. The problem experienced by these little ants is thus the opposite faced by that of the ponerines in dry woodland. Too much surface moisture can impede the development of young ants. Eggs and larvae can be kept on bare wet surfaces of the wood, but the pupae need a drier environment. The workers solve the problem by papering over some of the rooms and galleries with fragments of pupal cocoons from which adults have previously emerged. Sometimes the pieces are piled on top of one another to form several layers. The rooms have drier surfaces than others left bare, and the workers take care to move the pupae into them.

Nests located in the moist soil or in rotting wood are ideal growth chambers for countless bacteria and fungi that are potential health hazards for the ants. Nevertheless, ant colonies are rarely struck by bacterial or fungal infections. The reason for this remarkable immunity was discovered by Ulrich Maschwitz. He found that the metapleural glands in the thorax of adult ants continuously secrete substances that kill bacteria and fungi. Most remarkably, the fungus cultivated by the leafcutter ants *Atta* is not affected by the secretions, but all other foreign fungi or bacteria attempting to invade the *Atta* fungus garden are totally eliminated.

A queen of the miniature tropical American ant *Prionopelta amabilis*, surrounded by her daughter workers and cocoons containing both worker and queen pupae.

Ants as a whole have achieved a dominance across many land habitats enjoyed by few other groups of insects. Their numerical success has allowed them to alter not just their nest environments but the entire habitats in which they live. Harvesting ants, species that regularly include seeds in their diet, have an especially high impact. They consume a large percentage of the seeds produced by plants of many kinds in nearly all terrestrial habitats, from dense tropical forest to deserts. Their influence is not wholly negative. The mistakes they make by losing seeds along the way also disperse plants and compensate at least in part for the damage caused by their predation.

"Go to the ant thou sluggard; consider her ways": Solomon thus praised harvester ants for the industry they display in gathering seeds and their storage of the excess bounty in underground granaries. Writers of the ancient world were well aware of harvesting ants, because they lived in dry Mediterranean environments where the prudent habit is exceptionally well developed. The dominant species they encountered were most likely *Messor barbarus*, which occurs in the Mediterranean

A swarm raid by the army ant *Eciton burchelli* begins at dawn. In the background, under the fallen tree trunk, the massed bodies of hundreds of thousands of workers still shelter the queen and brood. Thousands more pour out of this bivouac to form a broad advancing front. In the foreground a group has overwhelmed a large whip scorpion; nearby a bicolored antbird *(upper left)* and barred woodcreeper *(upper right)* watch for insects flushed by the ants. (Painting by John D. Dawson, courtesy of the National Geographic Society.)

During a swarm raid, workers of the Asian army ant *Pheidologeton diversus* are supported by a gigantic supermajor, which works like a bulldozer to move obstacles out of the raid path. Members of the caste also crush prey between their mandibles. (Photograph by Mark Moffett.)

Above: in a Costa Rican forest, a worker of the extreme trap-jawed ant *Acanthognathus teledectus* stalks a springtail; when she touches the insect with her antennae, her widely opened mandibles snap shut. *Below:* the worker carries her impaled prey back to the nest. (Photographs by Mark Moffett.)

The "masters of camouflage" in the ant world are members of the tropical American genus *Basiceros. Above:* a portion of a *Basiceros manni* colony from Costa Rica, including earth-coated workers and larvae. *Below:* the workers are covered with specialized hairs that capture and hold fine soil particles, rendering them almost invisible against the forest floor on which they live.

Golden-haired *Polyrhachis* ants from Africa. Their conspicuous color may advertise their formidable weapons, which include hooklike spines that rise from their waists.

A reconstruction, with imaginary details in the surrounding vegetation and animals, of the single known colony of *Pheidole nasutoides,* found at La Selva, Costa Rica. The nest of the ants has been disturbed (in this hypothetical scenario) by a *Dendrobates* frog, a species known to feed on ants. Both majors and minors swarm out and run swiftly in erratic looping paths over the vegetation. This motion, together with their unique color pattern, causes the large-headed major workers to resemble nasute soldiers of the termite genus *Nasutitermes.* Several of the termite soldiers, on a foraging expedition, pause on a leaf to the left. (Painting by Katherine Brown-Wing.)

A mound of the wood ant *Formica polyctena* in the German forest. In the foreground, workers kill a sawfly larva, only one of some 100,000 prey items taken during a typical day. The structure of the mound increases the rate of warming in early spring, allowing the ants a head start over many of their competitors. (Painting by John D. Dawson, courtesy of the National Geographic Society.)

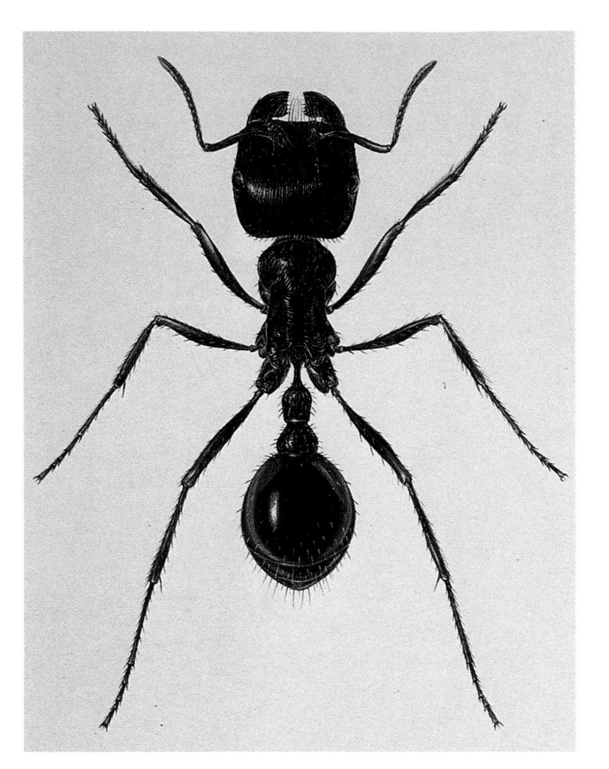

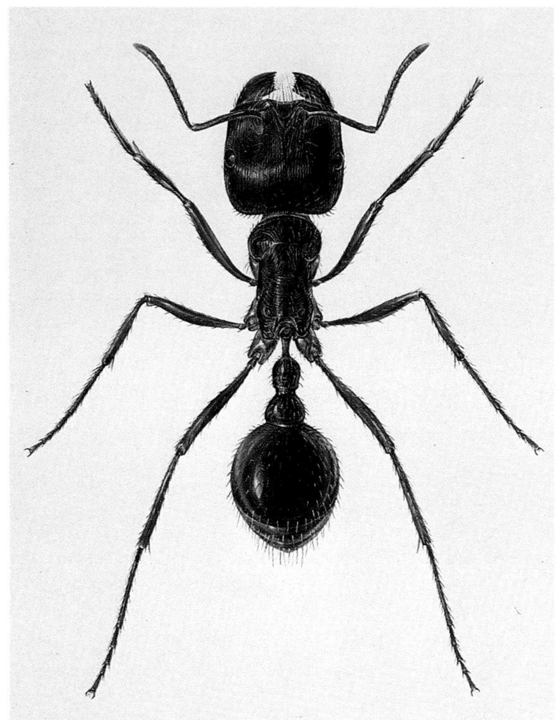

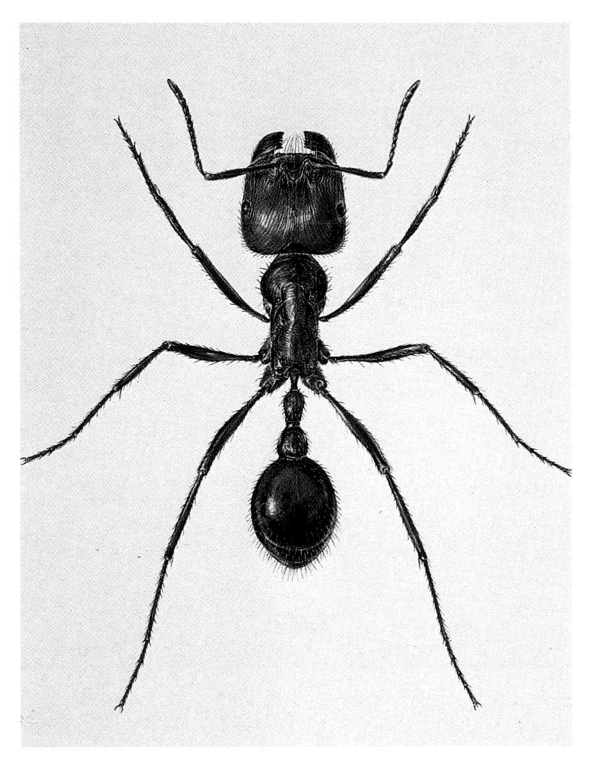

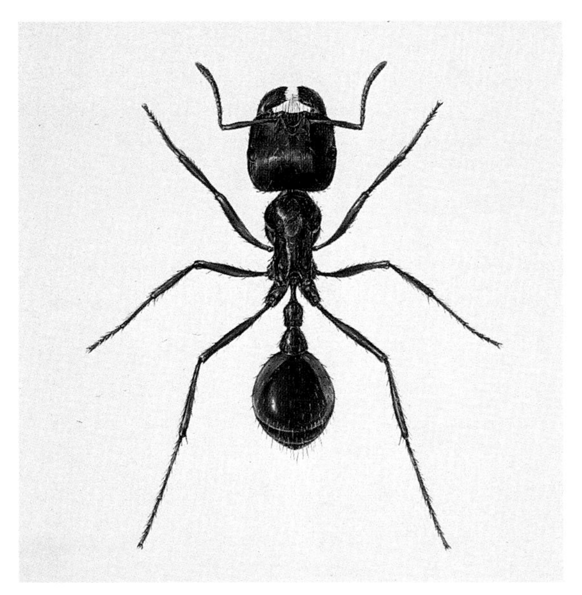

A gallery of North American *Pogonomyrmex* harvester ants (shown to scale).
Top left: P. rugosus. Top right: P. barbatus.
Bottom left: P. maricopa. Bottom right: P. desertorum.
(Paintings by Turid Forsyth.)

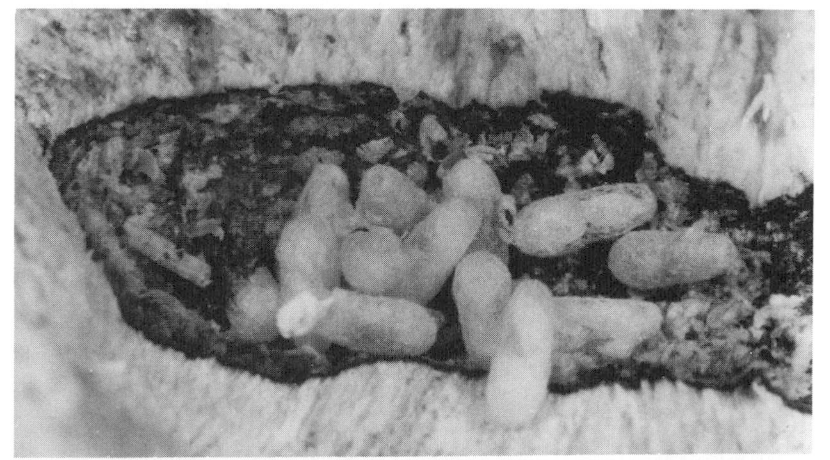

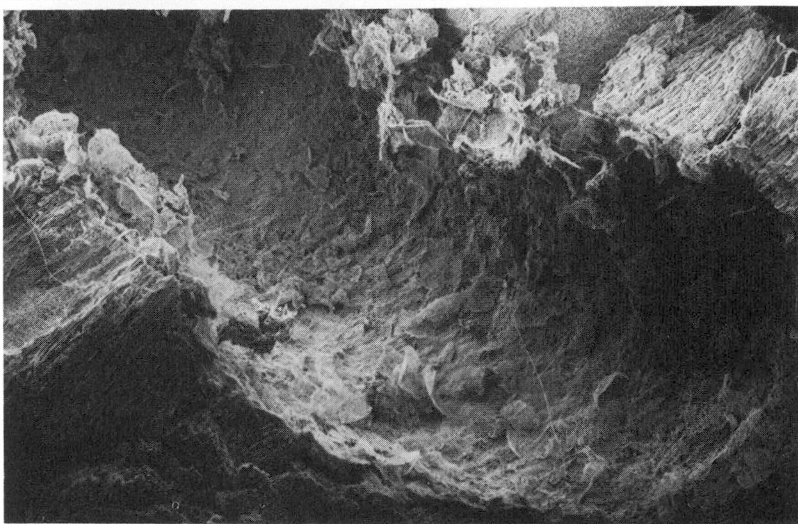

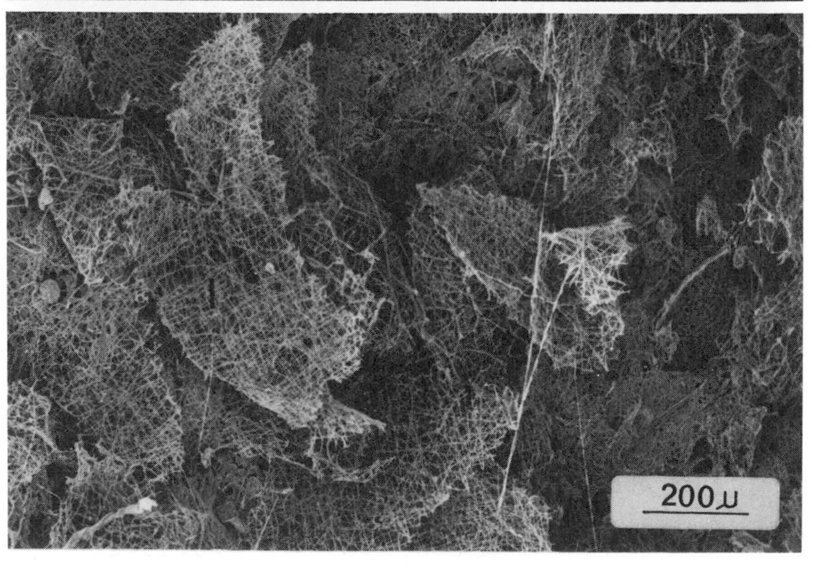

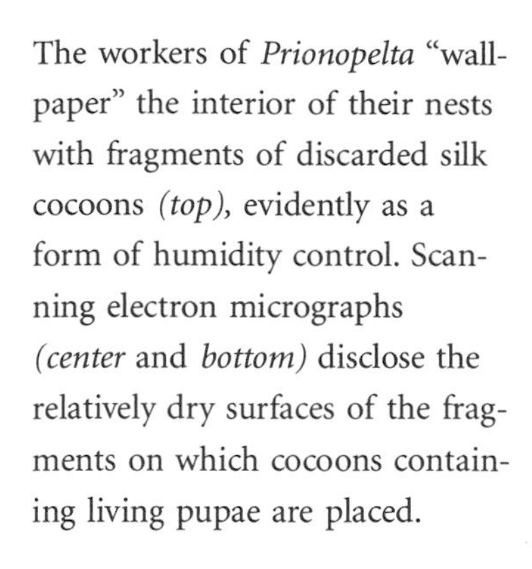

The workers of *Prionopelta* "wallpaper" the interior of their nests with fragments of discarded silk cocoons *(top)*, evidently as a form of humidity control. Scanning electron micrographs *(center* and *bottom)* disclose the relatively dry surfaces of the fragments on which cocoons containing living pupae are placed.

region and south through Africa; *Messor structor,* which is absent in Africa, but ranges all the way from southern Europe to Java; and *Messor arenarius,* which is abundant in the deserts of North Africa and the Middle East. These middle-sized, conspicuous ants are often serious grain pests, and it is to them that Solomon, Hesiod, Aesop, Plutarch, Horace, Virgil, Ovid, and Pliny almost certainly allude.

The first scientific observers of ants in the modern era, from the early 1600s to the early 1800s, were skeptical of the classical accounts despite the long list of authors who repeated the claim. And justifiably so, since without exception their experience was limited to northern Europe, one of the few parts of the world where the phenomenon is rare to non-existent. When European naturalists paid closer attention to ants in warmer, drier climates, the activity was reconfirmed. During a sojourn in southern France in the early 1870s, Reverend J. Traherne Moggridge, an American entomologist, explored seed harvesting by *Messor barbarus* and *Messor structor* in detail, establishing that the ants collect seeds from at least 18 families of plants. He confirmed the reports by Plutarch and other classical authors that the workers bite off the radicle to prevent germination, then store the deactivated seeds in granary chambers in the nests. In a remarkably modern addendum, Moggridge went on to prove that harvesters play an important role in dispersing plants by accidentally abandoning viable seeds in the nest vicinity or failing to deactivate them before they sprout inside the nest chambers.

In the last century exacting studies by biologists after Moggridge have touched on every aspect of the natural history of harvesters almost everywhere they occur, from Eurasia, Africa, and Australia to North and South America. One important finding is that the ants strongly alter the abundance and local distribution of flowering plants. They are especially potent in deserts, grasslands, and other arid habitats where harvesting is most intensive. They tip the balance in competition among some species of plants while promoting a balance in numbers among others. In so doing, they rearrange the distribution of species in local floras.

Harvesting by ants reduces the vegetative mass of plants as well as their reproductive power. Experiments performed in Arizona by James Brown and other ecologists revealed that when ants are removed from desert plots, annual plants grow in at double the ordinary density within just two seasons. In similar experiments performed in Australia by Alan Andersen, seedling numbers increased fifteen-fold.

Harvesting ants also often aid the exploited species by dispersing them more widely than would otherwise be the case. In the Arizona deserts, many seeds survive long enough to take root in the refuse piles around harvester nests, and thus certain plant species are scattered across the barren land from one nest site to another. These plants and the harvesters can be said to exist in a loose form of symbiosis. The plants "pay" the ants a certain fraction of their seeds in return for transport of another fraction of their seeds to the nest perimeters, which are rich in nutrients and nearly free of competitors.

By means of this unintentional manipulation, harvester ants exert powerful effects on the life and death of certain plants. They are keystone species, deciding by their presence alone which of the plants flourish and which fail. In croplands of the Mexican tropical lowlands fire ants *(Solenopsis geminata)* reduce the abundance of weeds among the domestic plants; they also cut the number of species of insects on the plants to one-third. The ants prefer some kinds of seeds over others. As a result, a few plant species are lifted to dominance while their competitors are driven to extinction. In other cases an equilibrium is reached. Plants that would otherwise eliminate competitors are cropped to low enough levels by the ants so that all coexist indefinitely.

Harvesting, with its unintended consequences, is only one of many symbioses that have existed between ants and plants for as long as tens of millions of years. By the middle of the Cretaceous Period, when dinosaurs still reigned, primitive sphecomyrmine and ponerine ants were on the rise; at the same time the flowering plants were diversifying and spreading around the world as the newly dominant form of vegetation. An intricate coevolution of the plants and insects as a whole was under way. Many of the plant species had come to depend on moths, beetles, wasps, and other insects for pollination, while an even greater number of insect species subsisted on nectar and pollen obtained during the pollination process. Another legion of insects fed on the foliage and wood of the flowering plants. The plants responded by evolving various combinations of thick cuticles, dense spines and hairs, and chemical defense substances such as alkaloids and terpenes, including chemicals that we humans now use in small doses as medicines, insect repellents, drugs, and condiments.

Into this lively theater of coevolution the ants entered. As the Cretaceous drew to a close, the ants increased in diversity and abundance, seized new roles as pollinators and seed dispersers, and appropriated parts of the plants as nest sites. An entomologist returning to early post-Cretaceous times, about 60 million years ago, would find familiar-looking ants swarming over familiar-looking vegetation.

Complex symbioses were fashioned among the thousands of species of ants and plants living together. The relationships found today are often parasitic, with ants exploiting plants and giving nothing in return. In other combinations they are commensalistic, with one partner making use of the other but, as in the case of ants occupying the dead hollow stems of trees and bushes, neither harming nor helping it. Of far greater general interest, however, are the mutualistic symbioses, from which both partners benefit. Ants use cavities supplied by the plants for nest sites, as well as nectar and nutritive corpuscles for food. In return, they protect their plant hosts from herbivores, transport their seeds, and literally pot their roots with soil and nutrients. Some pairwise combinations of ants and plants have coevolved so that each is specialized to use the other's services. The pacts of mutualism have produced some of the strangest and most elaborate evolutionary trends found in nature.

The classic case of complete interdependence is the symbiosis between members of the woody genus *Acacia* in Africa and tropical America and the ants living in them. Among the combinations, the American bull's-horn acacias and their ants have been the most thoroughly documented. The acacias, which are among the dominant shrubs and trees in dry forests, seem thoroughly designed to shelter and feed ants. Their thick pairs of thorns (the "bull's horns") are distributed at regular intervals up and down the branches. They are hard-shelled and inflated, and their pulp-filled centers serve as ideal shelters for ants. Nectaries exuding sugary liquid are located at the base of the feathery compound leaves. The workers need only step out of the entrance holes they have cut into the thorns and run a few centimeters in order to drink from droplets of the nectar. To these amenities the acacias add nutritious little buttons that sprout from the tips of the leaflets. The corpuscles, called Beltian bodies, are easily plucked off by the ants. All the evidence suggests that the dominant inhabitants of the acacias—slender stinging ants of the genus *Pseudomyrmex*—are able to thrive on nectar and Beltian bodies alone.

In return the ants protect the acacias from their enemies. They are crucial to the high success, indeed the very survival, of the plants. This side of the symbiosis was proven in a field experiment performed in the early 1960s by the American ecologist Daniel Janzen. During the course of studies in Mexico, Janzen, then a young graduate student, noticed that acacia shrubs and trees lacking *Pseudomyrmex* ants suffered greater damage from insects. They were also partly overgrown by competing plant species. When Janzen removed ants from occupied trees, by spraying with insecticides or clipping off the branches and thorns in which the *Pseudomyrmex* lived, he found that the acacias came under heavy attack by their insect enemies. Coreid bugs and treehoppers sucked on the shoot tips and new leaves; scarabs, leaf beetles, and caterpillars of assorted moths browsed on the leaves; and buprestid beetle larvae girdled the shoots. Other plants grew in more closely and shaded the stunted shoots.

In nearby occupied trees, those left untouched by Janzen, the ants attacked the invading insects, driving off or killing the great majority. Alien plants that sprouted within a radius of 40 centimeters of the acacia trunks were chewed and mauled by the ants until they died. Up to a fourth of the entire worker force of the occupied trees was active on the plant surfaces at any given time, day and night, constantly patrolling and cleaning the surfaces.

As Janzen's experiment proceeded, the ant-occupied trees thrived, while the empty trees progressively declined. In 1874 Thomas Belt, the naturalist who first recorded the symbiosis, had concluded that the *Pseudomyrmex* ants "are really kept by the acacia as a standing army." This view has now been firmly proved.

Similar ant-plant symbioses abound in tropical forests and savannas

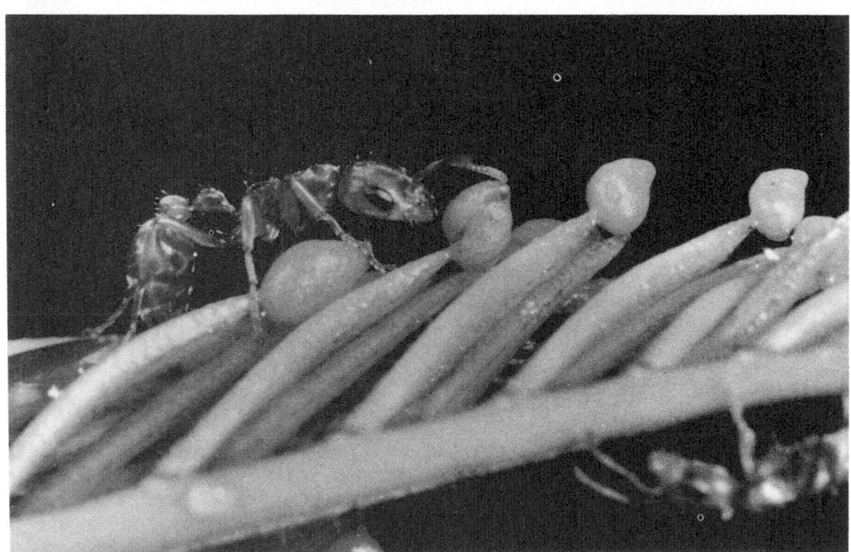

The bull's-horn acacias of tropical America harbor ants of the genus *Pseudomyrmex* in a close symbiotic relationship. In the upper photograph the nest entrance used by the ants is shown; in the foreground can be seen a row of nipplelike extrafloral nectaries from which the ants feed. In the lower photograph a worker collects nutritious Beltian bodies from the tips of the acacia leaflets. (Photographs by Dan Perlman.)

around the world. They have been the subject of an explosion of research in recent years. Ulrich Maschwitz and his collaborators, for example, have discovered a string of new symbioses in the rain forests of Malaysia, linking surprising new combinations of ant and plant species. Similar reports are coming in from Africa and both Central and South America. At the present time we know hundreds of plant species in more than 40 families that possess special structures to house ants. Many also supply nectar and food bodies, in the acacia manner. Among them are the legumes (including acacias), euphorbs, madders, melastomes, and orchids. The ants that depend on the symbiosis to some degree are equally diverse, including hundreds of species in 5 subfamilies.

Ants that are completely dependent on symbiotic plants are also among the most aggressive in the world. Those large enough to attack mammals, including human beings, are well-armed, quick and vicious. It is as though they have no other place to go, and with their backs to the wall they are prepared to make an extreme response to every provocation. The acacia ants swarm out almost instantly to mount and sting an offending arm and hand. When a person stands upwind and close by an acacia bush, some of the workers run to the edge of the leaves and strain to reach him, apparently aroused by his body odor alone. Larger and even more aggressive *Pseudomyrmex* ants inhabit *Tachygalia*, a small understory tree of South American forests. To brush one's bare skin against a sprig of *Tachygalia* is like touching a nettle. The punishment in this case, however, is delivered by dozens of ants that sprint onto the body, instantly begin to sting, and hold on until picked off. As we have walked through the rain forest undergrowth distracted, often rather carelessly in the typical naturalist's manner, we have felt the familiar burning sensation on some exposed part of our body and thought immediately: *Tachygalia!*

But the most effectively aggressive ant species in the world, exceeding even the tachygaliaphilic *Pseudomyrmex*, may be *Camponotus femoratus*, a large, hairy and decidedly unpleasant ant of South American rain forests. When disturbed in the slightest, the workers boil out in an angry mass over the nest surface. Just the close presence of a human being is enough to trigger the reaction. Diane Davidson, an American entomologist who has studied ant-plant symbiosis extensively, described the behavior in a letter to us as follows: "When I approached to within 1–2 m of their nests, workers of this species typically began to run back and forth and frequently jumped or fell onto me. Workers of all size classes of this polymorphic species attempted to bite, but usually only the major castes were capable of breaking the skin with their mandibles and causing a stinging sensation by simultaneously biting and spraying formic acid into the wound."

These ants happen to live not in the cavities of plants, but in ant gardens, which constitute the most complex and sophisticated of all symbioses between ants and flowering plants. The gardens are round masses of soil, detritus, and chewed vegetable fibers assembled in the branches of bushes and trees, ranging in size from golf balls to soccer balls, within which are grown a variety of herbaceous plants. The ants collect the materials for the nests. The ants gather the seeds of the symbionts and place them in the nests. As the plants grow, nourished by the soil and other materials, their roots become part of the framework of the gardens. The ants in turn feed on the food bodies, fruit pulp, and nectar provided by the plants.

The ant gardens of Central and South America contain many plant species, representing at least 16 genera, that are found nowhere else. These specialized forms include arums such as *Philodendron,* bromeliads, figs, gesneriads, pipers, and even cacti.

The plants limited to the gardens appear to be full symbiotic organisms. The ants transport their seeds to favorable sites within the nests, including the brood chambers, at least in part because the ants find them attractive and may even confuse their odors with those emitted by their own larvae. Some of the attractants have been identified, including 6-methyl-methylsalicylate, benzothiazole, and a few phenyl derivatives and terpenes. The growth of the plants is promoted by the activities of the ants. The ants in turn are less fully committed to the gardens. The diet supplied them by the plants is not restrictive; all the known garden-ant species forage away from the gardens to collect other kinds of food. Yet the ants that engage the symbiosis, including the ferocious *Camponotus femoratus,* seem to know they have a good thing going. At least they act as though their lives depended on it.

Epilogue: Who Will Survive?

LOCKED INTO THEIR chemosensory world, ants are oblivious to human existence. They experience reality for the most part through sensory devices projecting from their hard exoskeletons in the form of hairs, pegs, and plates. Their strange tripartite brains process information received primarily from a space of only a few centimeters around their bodies. They are, furthermore, aware of no more than a few minutes and hours of time into the past, and they have no mental construct of the future. It has been thus for tens of millions of years in the past, and so it will continue into the indefinite future. That difference in scale can never be abolished for a tiny creature imprisoned within an exoskeleton.

Because ants exist in a fractal world of centimeters, they are part of what human beings can profitably view as microwildernesses. Each colony grows and reproduces in a habitat contained by as little as two buttresses of a tree, the bark of a fallen log, or the soil beneath a scattering of stones. "Real" wilderness as human beings consider it, viewed over distances of hundreds of kilometers (again, a matter of perceptual scale), is everywhere threatened. Most of the forests and grasslands may vanish or be corroded almost beyond recognition, but some ant colonies will persist somewhere, and they will continue to cycle through their hereditary routines as though they were in a pristine world before the arrival of humankind. The superorganisms make no concessions, understand no mercy or variance given on their behalf, and will always be as elegant and pitiless as we now witness them, until the last one dies. But we are unlikely to see that happen. Their microwildernesses will outlast our own human-scale ecosystems.

Ants have lived on Earth for more than ten million of their generations; we have existed for no more than a hundred thousand human generations. They have evolved hardly at all during the past two million years. In the structure of our brain, we have undergone in the same period of time the most complex and rapid anatomical transformation in the history of life. Like a secondary rocket catching fire, our cultural evolution has accelerated change still more in a span of several centuries, exceeding the rate of organic evolution by orders of magnitude. We are the first species to become a geophysical force, altering and demolishing ecosystems and perturbing the global climate itself. Life would never die through the actions of ants or of any other wild creatures, no matter how dominant they became. Humanity, in contrast, is destroying a large part of the biomass and diversity of life, a success that perversely measures our own biological dominance.

If all of humanity were to disappear, the remainder of life would spring back and flourish. The mass extinctions now under way would cease, the damaged ecosystems heal and expand outward. If all the ants somehow disappeared, the effect would be exactly the opposite, and catastrophic. Species extinction would increase even more over the present rate, and the land ecosystems would shrivel more rapidly as the considerable services provided by these insects were pulled away.

Humanity will in fact live on, and so will the ants. But humankind's actions are impoverishing the earth; we are obliterating vast numbers of species and rendering the biosphere a far less beautiful and interesting place for human occupancy. The damage can be fully repaired by evolution only after millions of years, and only then if we let the ecosystems grow back. Meanwhile let us not despise the lowly ants, but honor them. For a while longer at least, they will help to hold the world in balance to our liking, and they will serve as a reminder of what a wonderful place it was when first we arrived.

How to Study Ants

WE WILL NOW present a primer of simple techniques for studying ants for students and for the diverse population of field researchers who need to handle material quickly and efficiently. Our exposition will be far from an exhaustive account. In the case of the culture of live colonies especially, specialized methods suited to the needs of particular species are often developed as part of research programs, and they can be found in the "Materials and Methods" sections of the respective technical articles. What we offer here is a set of general procedures that we have found to work well over our many years of experience, across almost all major groups of ants.

Collecting Ants

Collecting ants is simple and straightforward and anyone can do it very easily. We routinely place specimens in 80 percent ethanol or isopropyl alcohol; the latter substance is especially useful because it can be obtained as rubbing alcohol in many parts of the world without a prescription. (An unusual but workable approach was taken by the late astronomer and amateur myrmecologist Harlow Shapley, who used to preserve ants in the strongest spirits of the country he visited. A worker of *Lasius niger* he placed in vodka while dining with Stalin in the Kremlin is now in the Museum of Comparative Zoology at Harvard.) The vials we favor are small and slender, 55 millimeters long and 8 millimeters wide, dimensions that allow many to be kept in a small storage space or carried in a pocket or field pack. They are closed with neoprene stoppers, which permit preservation of "wet" material for many years. A few wider bottles, 55 millimeters long and 24 millimeters wide, are carried to accommodate the largest ants.

Workers should be collected whenever possible. You can mix them, as to both colonies and species, if you find the ants foraging singly (this fact should be noted on the label). If you discover the colony, however, you should put together a sample of at least 20 workers in a vial, along with up to 20 queens, 20 males, and 20 larvae if these can be captured. In emergencies, when the supply of vials is running low, you can place several nest series (that is, members of several colonies) in the same vial and separate them from one another with tight plugs of cotton. Up to four nest series can thus be accommodated in a typical, 55 × 8 millimeter vial. In small but clear letters, write a label of the following kind with a sharp pencil or indelible ink:

FLORIDA: Andytown, Broward Co.
VII-16–87. E. O. Wilson. Scrub hammock,
nesting in rotting palm trunk.

To pick up the ants, use stiff narrow forceps with pointed (but not needle-sharp) tips. A pair of very sharp watchmaker's forceps, for example Dumont No. 5, can be carried for use with exceptionally small ants. A rapid, efficient method is to moisten the tip of the forceps with alcohol from the vial and touch it to the ant; this procedure fastens the specimen to the forceps long enough to transfer it to the liquid in the vial. Fine, flexible forceps can also be carried for the collection of live specimens, if these are needed for behavioral observation.

To conduct a general survey of a particular locality, continue collecting until you have encountered no previously unseen species for a period of several days. Work primarily during the day, but search through the same area at night with a flashlight or headlamp to pick up exclusively nocturnal foragers. A good collector can obtain a virtually complete list of the fauna in an average site of 1 hectare (about 2½ acres) within one to three days. Habitats with dense, complex vegetation, however, such as those in tropical rain forests, are likely to take much longer and require special techniques such as arboreal fogging with insecticides.

For ordinary arboreal collecting, rake branches and leaves back and forth with a strong sweep net. Then break open hollow dead twigs on bushes and trees. This technique will reveal colonies of species, especially those with nocturnal habits, not readily discovered in any other way.

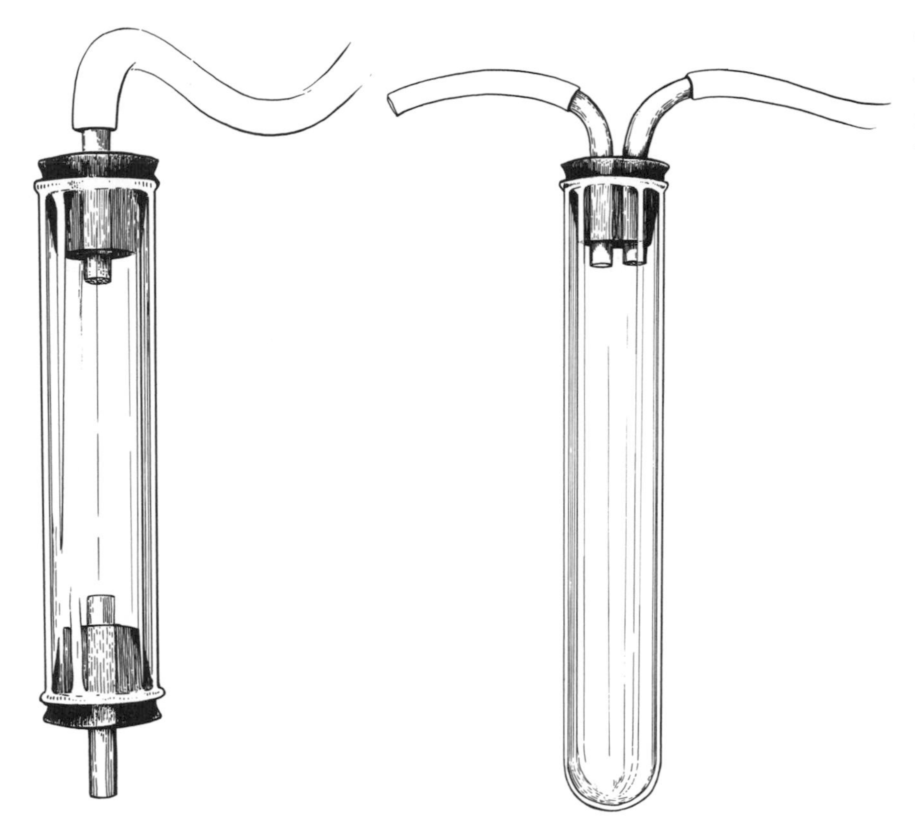

Two types of aspirators used in the rapid collection of ant specimens. The air tubes are covered by a wire or nylon mesh.

Often it is possible to make rapid, clean collections by snapping the inhabited twigs into short segments (3 to 6 mm long) and blowing the contents into the vial. An aspirator can also be used to suck up ants rapidly, particularly when the nest has just been broken open and the inhabitants are scattering. Exercise care in using this technique, because many ants produce large quantities of formic acid, terpenoids, and other volatile toxic substances. The unwary collector is in danger of contracting formicosis, a painful but not fatal irritation of the throat, bronchial passage, and lungs.

For terrestrial species, collect workers foraging on the ground both during the day and at night. It is necessary to look closely for certain species that are small and slow-moving and hence difficult to see. A favorite technique of ours in sampling forest faunas is to lie prone, clear the loose leaves from a square meter of ground to expose the soil and humus, then simply watch for up to a half-hour for the most inconspicuous ants. Another is to put out baits of pieces of tuna or cake crumbs and track the laden workers back to their nests.

In open terrain look for crater nests and other excavations, and with a gardener's trowel dig down into them in search of colonies. Turn over rocks and pieces of rotting wood on the ground to seek the species specialized for nesting in such protected sites. Tear open rotting logs and stumps, looking with special care beneath the bark for the small, inconspicuous species that abound in this microhabitat. Spread a ground cloth (a sheet of white cloth or plastic, 1 to 2 m on the side), and scatter leaf litter, humus, and topsoil over it. Break up rotting twigs and small tree branches buried within the litter. Where the humus and litter are relatively thick and moist, they often harbor a large part of the ant fauna and contain many inconspicuous and still poorly studied species.

The following technique has proved effective for collecting whole colonies that nest in small rotting logs and branches lying on the ground. Pick up a fragment of the decaying wood (say about 50 cm long), hold it above a photographer's developing pan or similar shallow-walled container, and strike the fragment with a trowel several times to shake out portions of the colony. Small pieces of the wood will also fall into the pan, but it is still much easier to locate and collect the ants, including entire colonies, this way than by ordinary excavation.

Slower but more thorough collecting of terrestrial ants can be accomplished with the aid of Berlese-Tullgren funnels, named after the Italian entomologist A. Berlese, who invented them and the Swede A. Tullgren, who modified and improved them. In simplest form the apparatus consists of a funnel topped by a wire-mesh screen onto which soil and litter are placed. As the material dries out, possibly aided by a light bulb or some other heat source placed above, the ants and other arthropods fall or slide down the smooth funnel sides into a collecting bottle partly filled with alcohol and suspended tightly under the lower spout of the funnel.

Preparing Specimens for Museum Work

Ants can be stored indefinitely in alcohol, but it is best to prepare part of the nest series as pinned, dried specimens for convenient museum work. This step is especially important if the ants are to be given to a taxonomist for identification. It is also the best way to store them in museums as voucher specimens, to serve as references for field or laboratory research (all such studies should be taxonomically verifiable with voucher material). The standard method for preparing dried specimens is to glue each ant on the tip of a thin triangle of stiff white paper. The tip should approach the right side of the ant and touch her ventral body surface beneath the coxae of the middle legs and hindlegs. The droplet of glue should be small enough and placed so as not to obscure any other part of the body except a portion of the coxae and ventral alitrunk surface—which have relatively few features of taxonomic importance. Prior to this "pointing" procedure, an insect pin should be inserted through the broad ends of two or three of the paper triangles, so that two or three ants from the same colony can be mounted one to a triangle on each pin. A rectangular label with the locality data goes beneath the mounted ants, so that when you read the label, the triangles point to the left and the ants point away from you. An effort should be made to get a maximum diversity of castes on each pin: for example, queen, worker, and male, or major worker, media worker, and minor worker. In the case of large ants, it may be possible to mount only one or two ants to a pin; and in the case of *very* large ants, it is sometimes best simply to pass the insect pin directly through the center of the thorax.

Culturing Ants

The culture and study of ants in the laboratory is a relatively simple operation. For many years we have used an economical arrangement that serves for both mass culturing and behavioral observation of a majority of species. The newly collected colony is brought into the laboratory (preferably with the queen and some of the original nest material) and placed in plastic tubs of a size to accommodate the size of the ants and the number of workers in the colony. For example, fire-ant colonies (*Solenopsis* species) with populations up to 20,000 are readily maintained in tubs about 50 centimeters long, 25 centimeters wide, and 15 centimeters deep. In order to prevent the ants from escaping, we use various means depending on the humidity of the room in which the ants are to be kept. The sides of the tub are coated with petroleum jelly, heavy mineral oil, talcum powder, or, preferably, Fluon (Northeast Chemical Co., Woonsocket, Rhode Island), a water-based material that is both effective (providing a silky smooth surface) and long-lasting (but unsatisfactory under humid conditions). The colony is allowed to settle into test tubes (15 cm long with inner diameters of 2.2 cm) into which water has been poured and then trapped at the bottom with tight cotton plugs, leaving about 10 centimeters of free air space from the plug to the mouth of the tube. The 10-centimeter segment is surrounded by aluminum foil to darken the air space and encourage the ants to move in (most do so promptly). It can be removed later to allow behavioral studies; most ant species adapt well to light at ordinary room intensities, carrying on brood care, food exchange, and other social activities in an apparently normal manner. The tubes are stacked at one end of the tub prior to insertion of the colony, leaving most of the bottom surface of the tub bare to serve as a foraging arena.

The nest tubes can also be placed in closed plastic boxes, making it easier to keep the ambient air of the foraging arena moist and hence better suited for forest-dwelling species. The following dimensions are roughly correct for ant species with different-sized workers:

Small. 11 × 8.5 centimeters on the side and 6.2 centimeters deep. Very small ants such as *Adelomyrmex, Cardiocondyla, Leptothorax,* small *Pheidole,* and *Strumigenys.* These species can also be cultured readily in small, round petri dishes (10 cm diameter, 1.5 cm depth).

Medium. 17 × 12 centimeters on the side and 6.2 centimeters deep. For example, *Aphaenogaster, Dorymyrmex,* and *Formica.* Smaller colonies of *Camponotus, Messor,* and *Pogonomyrmex.*

Large. 45 × 22 centimeters on the side and 10 centimeters deep. For example, larger colonies of *Pheidole, Pogonomyrmex,* and *Solenopsis.*

Variations on the elementary test-tube arrangement can be adapted to ant species with unusual nesting habits. Colonies of arboreal stem-dwelling ants such as *Pseudomyrmex* and *Zacryptocerus* can be induced to move into glass tubes 10 centimeters long with diameters of 2–4 millimeters, the latter dimensions varied according to the size of the workers. The tubes are closed at one end with cotton plugs. The plugs can be kept moist, but in many cases this is not necessary, because stem-dwelling ants are often adapted to dry nest interiors, and a small dish of water placed nearby is an adequate source of moisture. Each set of tubes containing a colony is then placed in a tub of the kind just described. Or the tubes can be bound horizontally in rows on a rack or potted plant, in order to simulate the natural environment.

Colonies of small fungus-growing ants can be maintained easily in moistened tubes in tubs. Large fungus growers, such as leafcutter ants of the genera *Acromyrmex* and *Atta,* are better kept via a technique developed by the American entomologist Neal Weber. Newly inseminated queens or incipient colonies are collected in the field and transferred to a series of closed, clear plastic chambers each about 20 × 15 centimeters and 10 centimeters deep (ordinary refrigerator food receptacles with transparent sides serve very well). The chambers are connected by glass or plastic tubes 2.5 centimeters in diameter, allowing the ants to move readily from one chamber to the other. Foraging workers are permitted to collect fresh vegetation (possibly supplemented by dry cereal) from empty chambers, from an open tub whose walls are lined with Fluon, or from a tub surrounded by a moat containing water or mineral oil. As the colony expands in size, the ants fill one chamber after another with the characteristic spongelike masses of processed substratum, through which the symbiotic fungus grows luxuriantly. Except in the driest laboratory environments, no special water supply is needed, because the ants obtain all the moisture required from the vegetation. Leaves from a wide range of plant species are accepted by the ants. In the northeastern United States we have most frequently used basswood (linden), oak, maple, and lilac; the last two are especially attractive to the foragers. The colonies deposit the exhausted substrate in some of the chambers, which can be removed and cleaned from time to time.

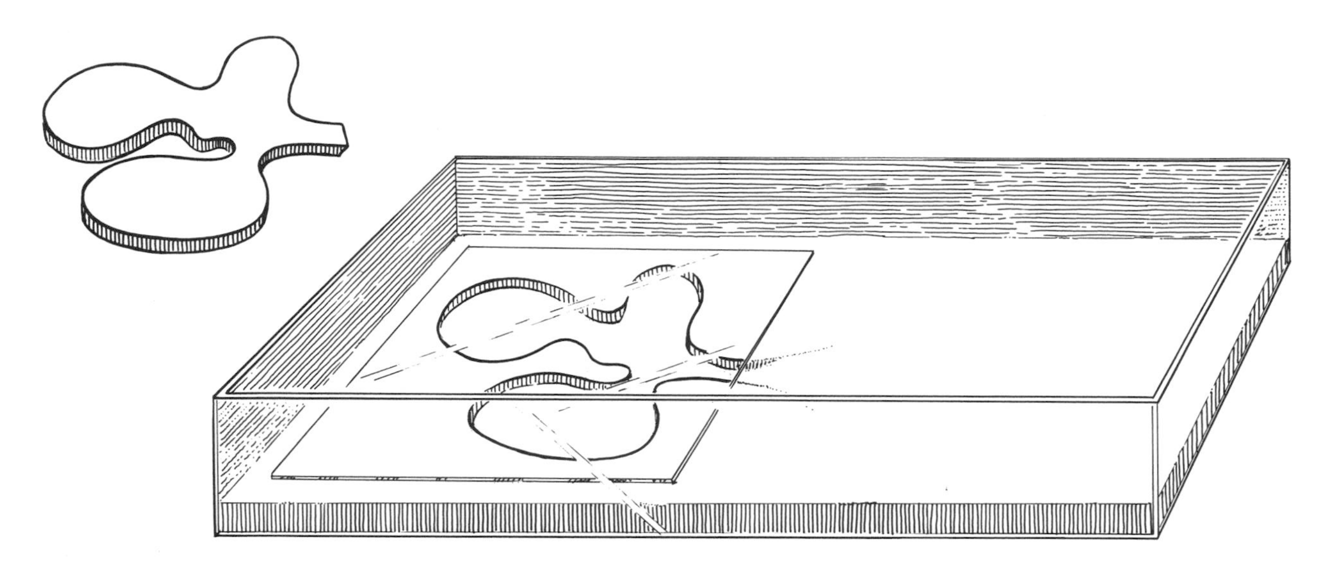

A method for making artificial nests. A template *(left)* imitating the ant nest in the field is made from Plasticine or precision polyether duplicating material (Reprogum). It is then placed on the bottom of the arena, and plaster of Paris poured around it. Next, a glass plate is gently placed on top of it. When the plaster hardens, the plate is lifted and the template removed. The glass plate is then put back on top. The ants have access to the arena through the small tunnel to the right of the impression.

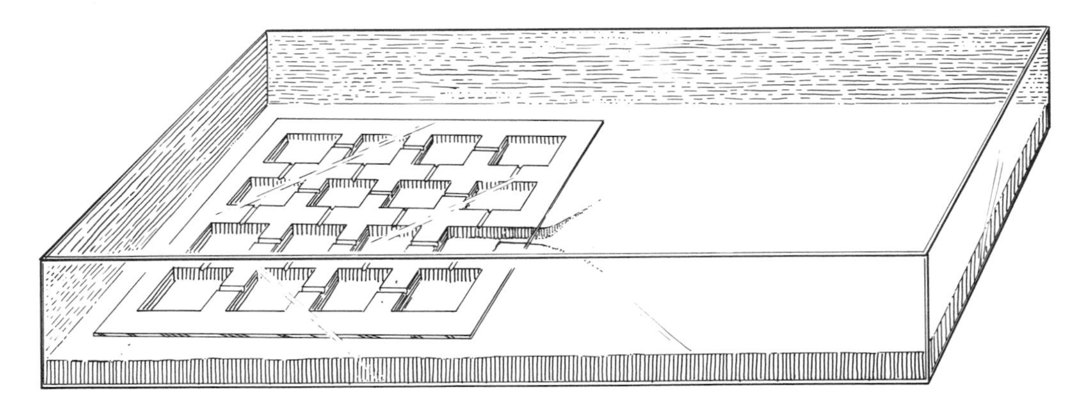

A horizontal nest with multiple chambers made from plaster of Paris. The nest sits in the plaster floor of a large foraging arena. The chambers are interconnected and covered by a glass plate. Pouring water around the glass plate at regular intervals keeps the nest chambers moist.

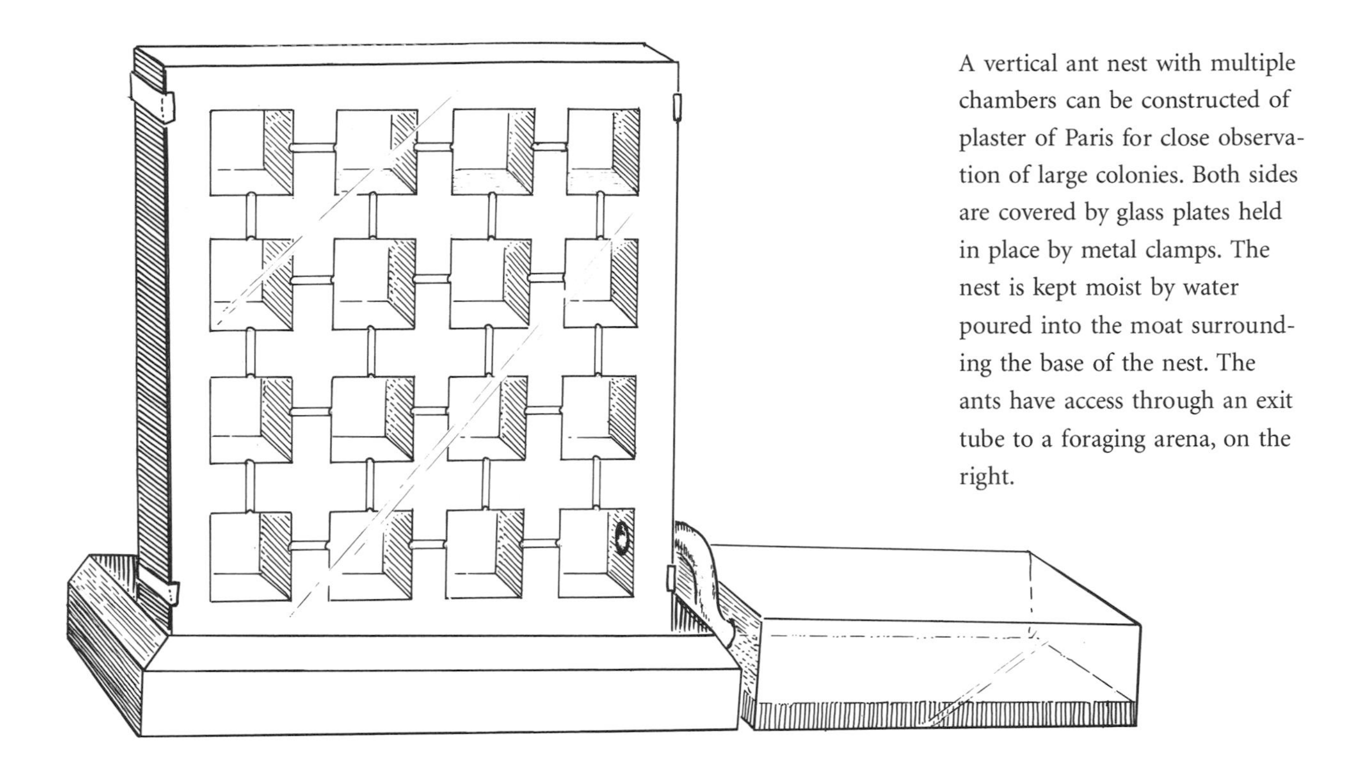

A vertical ant nest with multiple chambers can be constructed of plaster of Paris for close observation of large colonies. Both sides are covered by glass plates held in place by metal clamps. The nest is kept moist by water poured into the moat surrounding the base of the nest. The ants have access through an exit tube to a foraging arena, on the right.

For close behavioral studies, more elaborate artificial nests are often required. One that works well for the great majority of ant species can be prepared as follows. Pour plaster of Paris to a depth of 2 centimeters in the bottom of a tub whose size suits the worker size and population of the colony under study (for minute ants such as thief ants the container may be only 10 × 15 cm and 10 cm deep). As the plaster of Paris sets, carve into its surface 10 to 20 chambers that are roughly similar in size and proportion to the natural nest chambers of the colony to be cultured. In the case of some medium-sized ant species living in pieces of rotting wood, the chambers are typically ovoid or circular in shape and 1–4 centimeters across; hence chambers should be excavated that are about 2 × 3 centimeters wide and 1 centimeter deep. The artificial nest chambers are connected by galleries 5 millimeters wide and deep, and are covered tightly by a rectangular glass plate. Two to four exit galleries are cut from the outermost chambers to the remainder of the plaster of

Entire societies of *Leptothorax* and other small ants fit into nest cavities between two microscope slides (76 × 26 mm). The cavities are cut out from cardboard or plexiglass sheets and shaped to resemble the natural chambers. The nests are covered with a red foil to keep them dark to the ants' vision yet visible to human observers. If necessary, the filter-paper floor can be kept moist by adding drops of water.

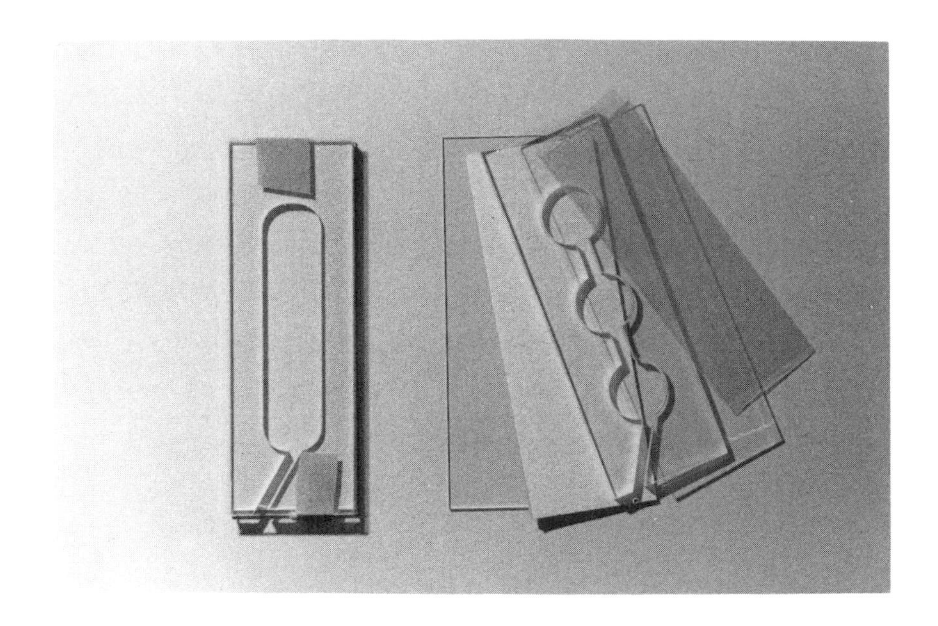

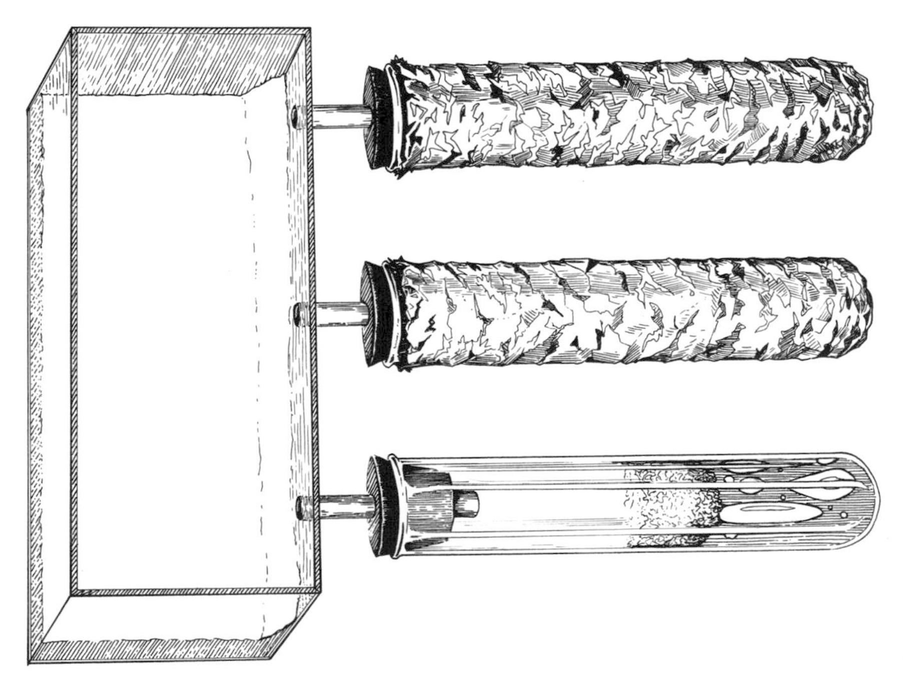

Glass test tubes, which can be covered with aluminum foil to darken their interiors, serve as readily constructed artificial nests that work well for many ant species and are easily carried on field excursions. The chambers are kept moist by water trapped behind tightly packed absorbent cotton balls, as shown in the lowermost tube. The chambers are connected to a foraging arena by narrow glass tubes inserted through neoprene stoppers.

Paris surface, which serves as the foraging arena. Fragments of decaying wood and leaves from the vicinity of the original nest can be scattered over the surface to add to the "naturalness" of the microenvironment.

To construct a large number of plaster of Paris nests, we use a mold fashioned from Plasticine or rubber whose surface is in the shape (in reverse) of the chambers and galleries. Liquid plaster of Paris is poured over this mold. When the gypsum hardens, it is pulled up to form the upper part or even all of the artificial nest.

To feed ants in the laboratory, we employ the Bhatkar diet (named after its inventor, Awinash Bhatkar), which is prepared as follows:

1 egg
62 ml honey
1 gm vitamins
1 gm minerals and salts
5 gm agar
500 ml water
Dissolve the agar in 250 ml boiling water. Let cool. With an egg beater mix 250 ml water, honey, vitamins, minerals, and the egg until smooth. Add to this mixture, stirring constantly, the agar solution. Pour into petri dishes (0.5-1 cm deep) to set. Store in the refrigerator. The recipe fills four 15-cm-diameter petri dishes, and is jellylike in consistency.

Most insectivorous ant species thrive on this diet when fed it three times weekly along with fragments of freshly killed insects, such as mealworms *(Tenebrio)*, cockroaches *(Nauphoeta)*, and crickets, offered in small quantities. If the ants are also predators, they do especially well when allowed access to bottles containing fruit flies, preferably flightless mutants. Alternatively, the fruit-fly adults can be frozen and sprinkled onto the foraging arenas for the ants to discover.

Transporting Colonies

Colonies can be kept for days or weeks at a time in bottles or other tight containers, provided that certain elementary procedures are followed.

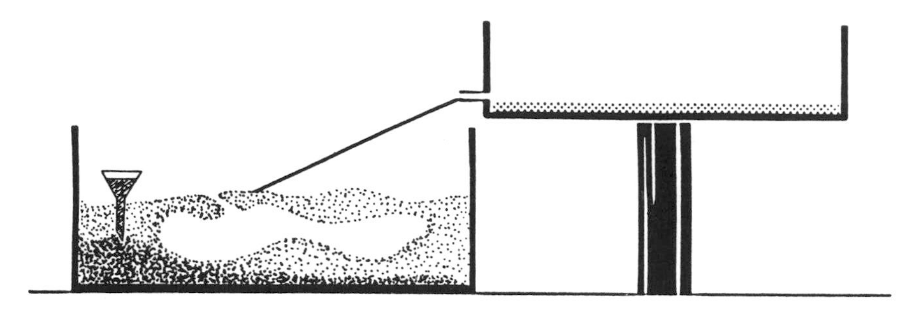

A sizable colony of harvester ants can be easily cultured in a glass terrarium partly filled with sand. The sand is kept moist, at least toward the bottom of the terrarium, by periodic watering through a funnel. The ants build their own nest chambers in the sand and forage across a thin wooden bridge to an arena, shown elevated on the right.

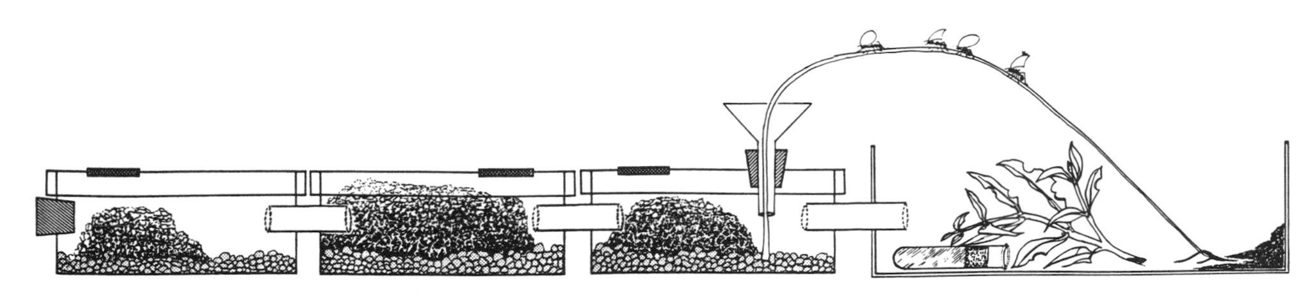

Leafcutter ants *(Atta),* despite their large and very complex societies, are easily cultured in the array of chambers illustrated here. The colony, including the mother queen, is housed in plastic boxes, each of which is approximately 15 × 20 × 10 centimeters in size. Clay pebbles may be placed on the bottom to help regulate humidity. The lids of the boxes are fitted with small openings covered by fine wire mesh, for improved ventilation. Several boxes can be connected by glass tubes when a small colony is first installed, and additional boxes can be added as the colony grows. One box opens through a funnel, which is coated inside with talcum powder to prevent ants from crawling over it. A flexible willow twig passes from the funnel to a foraging arena, whose walls are coated with Fluon or talcum powder to prevent escape. Leaves are placed in the arena for the ants to harvest, along with a water tube that provides extra moisture for the ants as needed.

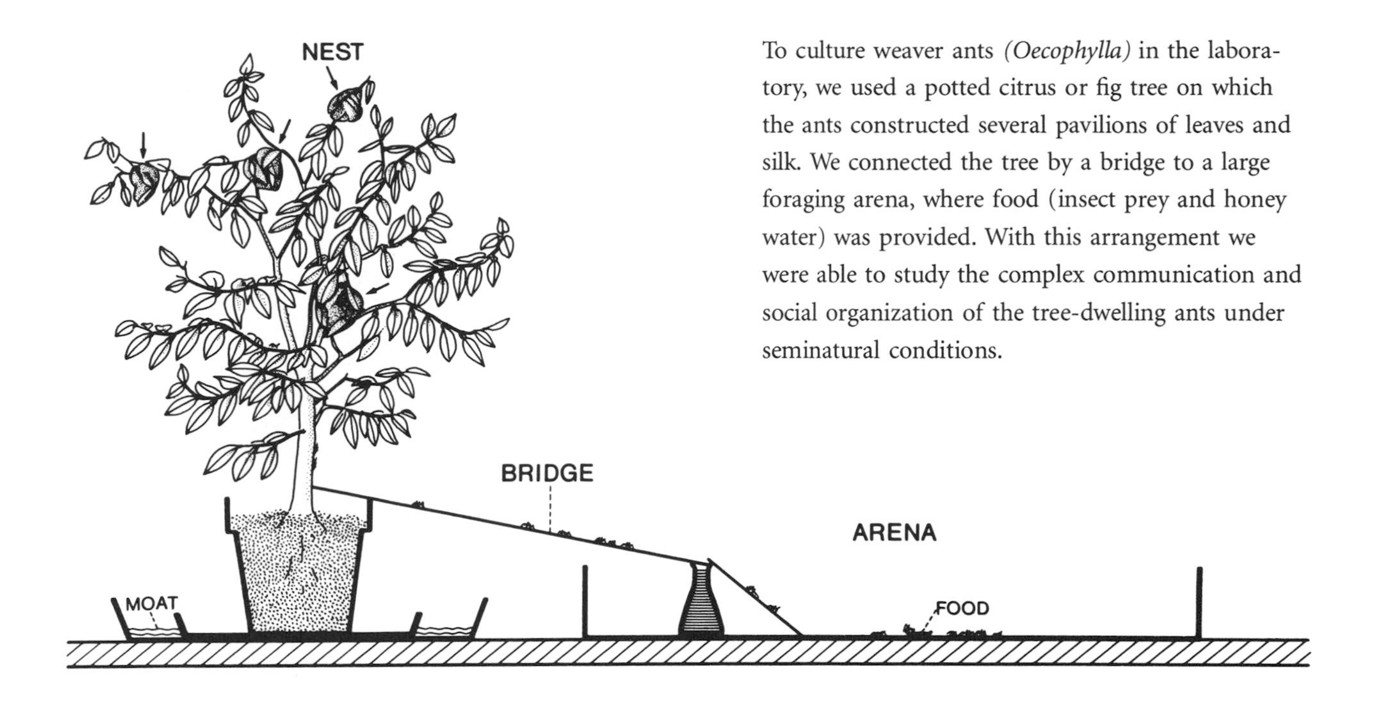

To culture weaver ants *(Oecophylla)* in the laboratory, we used a potted citrus or fig tree on which the ants constructed several pavilions of leaves and silk. We connected the tree by a bridge to a large foraging arena, where food (insect prey and honey water) was provided. With this arrangement we were able to study the complex communication and social organization of the tree-dwelling ants under seminatural conditions.

The first, absolute rule is that the ants must be given a moist area into which to retreat—not soaking wet, with films or drops of water that might entrap the ants, but an environment with a moist surface and saturated ambient air. The ideal retreat is part of the nest material itself, placed directly in the container, preferably with a portion of the colony in it. A large piece of moistened (but not dripping wet) cotton wool or paper toweling should be added as backup. The rest of the container can be filled with nest material or loose-fitting paper toweling or other neutral material to prevent the colony from being knocked about excessively in transport.

The colony should be uncrowded, in no case occupying more than 1 percent of the container volume. The lid of the container should be tightly fitted. Unless the colony is unusually active or aggressive, it is not necessary to punch holes in the lid to aerate the interior; in fact, this procedure risks excessive drying. Once or twice a day the lid can be removed and the container waved gently back and forth to freshen the

Weaver-ant colonies, as well as those of other arboreal ants, can be cultured in "test-tube trees," consisting of rows of the tubes mounted with clamps on laboratory stands *(right)*. The bottom quarter of each tube is filled with water, which is trapped in place by tight cotton plugs. In our cultures the ants closed the entrances and subdivided the living spaces by constructing walls of silk, as shown in the two views on the far right; the upper photograph shows a wall face on, and the lower shows several walls from the side.

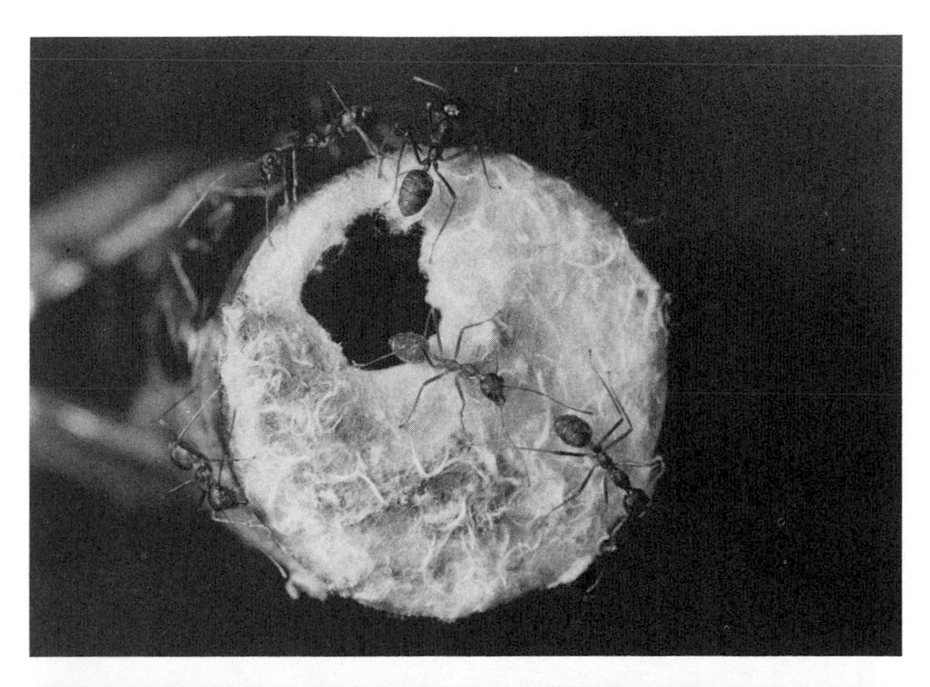

air. The colony can be given drops of sugar water and fragments of insects or other foods if the duration of the journey is more than several days. If ants appear dead after remaining in a closed container too long, they may just be narcotized by carbon dioxide. Expose them to the open air for a few hours to see if they can recover.

Because many countries have restrictions on the importation of live insects, it is prudent to check with the appropriate government agencies before collecting live colonies abroad. In the United States, for example, the U.S. Department of Agriculture (Animal and Plant Health Inspection Service, Plant Protection and Quarantine, Plant Importation and Technical Support) must furnish a permit, which has first to be approved by the appropriate state officials. The entire procedure usually requires six to eight weeks. The permit must be presented to the appropriate customs officer upon reentry into the United States.

An increasing number of countries restrict the export of preserved and living specimens, including insects, and a special export permit may be required. Local regulations should always be consulted and respected.

Acknowledgments

Unless otherwise indicated, all the illustrations belong to us. The ants *(Tetramorium caespitum)* lining the margins of the chapter openings are by Amy Bartlett Wright. The sources of works by others are acknowledged in the captions. We are especially grateful to the National Geographic Society for permission to reproduce some of the remarkable paintings by John D. Dawson which appeared in the article "The Wonderfully Diverse Ways of the Ant," by Bert Hölldobler, *National Geographic,* June 1984, pp. 778–813.

For their invaluable expert assistance, we are greatly indebted to Kathleen M. Horton (manuscript preparation and bibliographic research), Helga Heilmann (photographic processing), and Malu Obermayer (technical assistance).

Index